D0783205

NATIONAL AUDUBON SOCIETY POCKET GUIDE

NATIONAL AUDUBON SOCIETY

The mission of the NATIONAL AUDUBON SOCIETY *is to conserve and restore natural ecosystems, focusing on birds and other wildlife for the benefit of humanity and the Earth's biological diversity.*

We have 500,000 members and an extensive chapter network, plus a staff of scientists, lobbyists, lawyers, policy analysts, and educators. Through our sanctuaries we manage 150,000 acres of critical habitat.

Our award-winning *Audubon* magazine, sent to all members, carries outstanding articles and color photography on wildlife, nature, and the environment. We also publish *Audubon Field Notes,* a journal reporting seasonal bird sightings continent-wide; *Audubon Activist,* a newsjournal; and *Audubon Adventures,* a newsletter reaching 600,000 elementary school students. Our *World* Audubon television shows air on TBS and public telev

For information about how you can become a membe please write or call the Membership Department at:

NATIONAL AUDUBON SOCIETY
700 Broadway
New York, New York 10003
(212) 979-3000

GALAXIES AND OTHER DEEP-SKY OBJECTS

Text By Dr. Gary Mechler

Alfred A. Knopf, New York

This is a Borzoi Book
Published by Alfred A. Knopf, Inc.

Prepared and produced by Chanticleer Press, Inc.,
New York.
Typeset by Chanticleer Press, Inc., New York.
Printed and bound by Tien Wah Press, Singapore.

Published March 1995
First Printing

Library of Congress Catalog Card Number: 94-41623
ISBN: 0-679-77996-5

Contents

How to Use This Guide

Galaxies, nebulae, and star clusters, our fascinating and mysterious neighbors in the vast, deep space that surrounds our solar system, provide a wealth of information about the nature and formation of the universe, as well as fascinating viewing opportunities.

Coverage

This guide covers more than 70 deep-sky objects that are visible to the amateur skywatcher, are representative of a specific type or phenomenon, are particularly interesting, or have provided information about the universe.

Organization

This easy-to-use pocket guide is divided into three parts: introductory essays, illustrated descriptions of deep-sky objects, and appendices.

Introduction

The introductory essay "The Universe" explains how the universe formed billions of years ago. "Stars" discusses the evolution of these components of deep-sky objects. The different types of objects, including galaxies, quasars, nebulae, and star clusters, are described in "Deep-Sky Objects." Key to the Color Plates identifies the symbols and drawings used in the accounts.

The Deep-Sky Objects

This section includes 80 images and photographs of galaxies, nebulae, star clusters, and other phenomena of deep space, each accompanied by a detailed text description. The different types of galaxies are covered first, followed by interacting galaxies, galaxy clusters, and our own Milky Way. Nebulae are covered next, both those involved in the births of stars and those formed upon the deaths of stars, followed by open and globular star clusters. For each object we give the name or catalog number(s) by which it is known, the type of object it is, and the constellation in which it lies. Each text description is accompanied by an illustration of the type of object shown and a symbol indicating the minimum optical aid needed to see it, as well as its distance and magnitude, if applicable (see Key to the Color Plates).

Appendices   "Observing Deep-Sky Objects" offers information on skywatching and on viewing instruments. "Systems of Measurement" discusses systems used to locate and measure objects in the sky. The index lists all the objects covered in this book by their names and catalog designations.

The Universe

Although early civilizations studied and named the stars, it wasn't until after the invention of the telescope in the 1600s that human beings began to comprehend the vastness of the universe. Since then, with technological advances and individual contributions, our understanding of the universe and its origins has increased enormously.

Simply stated, the universe is everything with which we are physically connected, mainly by light and other forms of radiation. It includes our planet, solar system, and galaxy, billions of stars, star clusters, galaxies, and nebulae, and all the seemingly empty space around them. Stars and interstellar gas and dust, all held together by mutual gravitational forces, make up galaxies. Spaced thousands to millions of light-years apart, galaxies occur in clusters. Galaxy clusters are grouped in superclusters. Recent evidence points to even larger organizations, so vast they pose a serious challenge to our theories of how the universe evolved. One factor that complicates our study of the universe is that everything in it is in motion. Clusters of galaxies are moving apart from one another; galaxies move relative to one another; within galaxies, stars, star clusters, and nebulae orbit the galactic center.

Formation of the Universe	The current consensus (although it is not without its opponents) is that our universe began in an incredibly energetic explosion—the Big Bang—some 14 to 20 billion years ago. In the milliseconds that followed the explosion the universe was pure energy. Some of that energy immediately became matter, which condensed and eventually formed into gas, dust, stars, and galaxies. Most of the material in the universe is hydrogen gas, and much of the remainder is helium gas, while only a minute portion takes the form of elements heavier than helium. The material out of which we and all other objects in the universe are made (carbon, oxygen, nitrogen, and so on) is literally cosmic debris.

Although the Big Bang happened billions of years ago its aftereffects can still be detected. One consequence of the initial explosion is the ongoing expansion of the universe, which means that distant objects in the universe get farther away all the time. Because the distances are so great and the speed of light is finite, when we look out into space we are actually looking back in time, for it has taken the light we see from other galaxies millions to billions of years to reach us. We see these galaxies, then, as they appeared millions or billions of years ago. Evidence of the universe's formation—the now faint, cool radiation left over from the Big Bang—reaches us as weak radio waves, or cosmic background radiation (CBR), coming from all directions. It is very difficult to study the distant parts of the universe, however, because the CBR that reaches us is so faint, and some frequencies cannot even penetrate Earth's atmosphere.

Viewing the Universe

In ancient times human beings imagined that the stars in the night sky represented the shapes of mythological heroes, creatures, and objects. These patterns became known as constellations. In modern astronomy the term constellation refers to a particular region of the sky, enclosed by borders established in 1930 by the International Astronomical Union, who kept as close as possible to the ancient constellations. The concept of a constellation is simply a convenience, indicating only a direction of the sky toward which we can look to find a specific object, say a galaxy. From our point of view all the stars within a constellation seem to be physically related, but they may be many thousands of light-years apart.

Stars

Most deep-sky objects are made up of stars (galaxies, star clusters) or are associated with the births (nebulae) or deaths (supernova remnants) of stars. Stars form within clouds of gas and dust in space (nebulae) when random swirling motions, collisions between clouds, or the explosion of a nearby star cause the cloud to contract. As the cloud compresses it heats up, ultimately reaching several million degrees at its center. The new star produces energy by nuclear fusion, converting hydrogen to helium and yielding energy. Stars in this hydrogen-using stage, such as our Sun, are known as *main sequence stars.* At later stages stars fuse helium into carbon and oxygen, which in turn may fuse into heavier elements.

When an average-size star has exhausted the hydrogen fuel at its core it enters the *red giant* stage, expanding, cooling, and using up its store of helium. Ultimately it sheds its outer envelope, which becomes a planetary nebula, and its interior begins to shrink and heat up, becoming a *white dwarf,* an extremely hot, dense star. Very massive stars become *red supergiants,* which may become as large around as the orbit of the planet Jupiter. Such massive stars become unstable and variable in light output, and a very few become *supernovas,* exploding spectacularly and leaving behind a tiny, incredibly compressed core called a *neutron star* or *pulsar.* This small star usually rotates rapidly, sending out beams of light and radio waves. Some neutron stars may become *black holes,* regions so dense that even light cannot escape their intense gravitational fields.

Star Classification

Astronomers classify stars by spectral type, determined by the strengths and positions of absorption lines in their spectra. These absorption lines, though manifested as colors, are a function of temperature, so astronomers can determine a star's temperature from its spectrum. From hottest to coolest (that is, bluest to reddest) the major spectral types are O, B, A, F, G, K, M. Stars' spectral types offer clues to activity in the area around them. For example, the hottest, bluest stars are not long lived, so when scientists find these types of stars, they know there has been recent star formation in the area.

Most bright stars are assigned a Greek letter in the order of their brightness in their constellation, with the brightest star usually called alpha (α) and the second-brightest beta (β), and so on.

Deep-Sky Objects

Beyond our solar system lie intriguing objects composed of stars or gas and dust clouds or both. Collectively referred to as deep-sky objects, these include galaxies, star clusters, and nebulae. These objects have been cataloged by astronomers, including, in the late 1700s, Charles Messier (the "comet ferret"), who compiled a list of "objects to avoid" that frustrated his efforts to find comets. Later astronomers slightly modified and extended Messier's original list, and these 100-odd "Messier objects," mainly star clusters, galaxies, and nebulae, are among the most popular targets for stargazers. The New General Catalog (NGC) and Index Catalog (IC) of deep-sky objects were compiled at the end of the 19th century (since revised). Many deep-sky objects have both Messier (M) and NGC (or IC) numbers. In this book if an object has both, we have given the Messier number first, followed by the NGC or IC number.

Galaxies   Aggregates of gas, dust, and millions, billions, or trillions of stars held together by mutual gravitational forces, galaxies range from the comparatively small (dwarf) to the immense (giant). Common galaxies are divided into three basic categories by their shapes.

13

Elliptical galaxies, or type E galaxies, range in shape from essentially spherical (type E0) to highly elongated and flattened (type E7). Elliptical galaxies contain little or no gas and dust, and are thought to consist almost entirely of old stars. Most ellipticals are dwarf galaxies, but a small percentage, the giant ellipticals, are massive. Elliptical galaxies far outnumber *spiral galaxies* (type S galaxies), flattened disks with a central bulge (nucleus) from which curving arms extend. *Barred spiral galaxies* (type SB), with a bar of stars and interstellar matter running through their nuclei, are a subset of spiral galaxies. Spiral galaxies range from type S0, which are fairly featureless disks, through Sa (and SBa), with relatively large central bulges and tightly wound spiral arms, to Sb and Sc (and SBb and SBc), with smaller nuclei and increasingly spread-out spiral arms. The Milky Way is thought to be a type Sb or Sbc (intermediate between Sb and Sc). Spiral galaxies contain much gas and dust as well as both old and young stars. *Irregular galaxies* are usually small galaxies with no regular shape or only a hint of one. *Peculiar galaxies* are those that don't fit neatly into any of the other categories; they are often extremely energetic.

Quasars (quasi-stellar radio sources) are extremely bright, extremely distant high-energy objects. Because they are so distant, we see them as they appeared billions of years ago, when they and the universe were young. They may be the energetic cores of young galaxies or of colliding galaxies in the early, more crowded universe. Quasars are as intrinsically luminous as many galaxies combined, and they can emit enormous amounts of energy in the radio region of the electromagnetic spectrum, but their main energy source is extremely small. The most likely source of a quasar's power is a supermassive *black hole,* a region so dense that not even light can escape its intense gravitational field. The black hole pulls in stars, gas, and dust surrounding the quasar's center; as these are compressed and heated, they emit the radiation we perceive as a quasar.

Star Clusters   Many stars and star systems exist by themselves, but many others occur in clusters. Stars in a cluster are similar to one another in age and composition, having formed from the same interstellar cloud at about the same time. The stars in a cluster are gravitationally bound to one another and travel through space together.

There are two types of star clusters, open and globular. *Open clusters* (also called galactic clusters because they are found in the disk of a spiral galaxy) are relatively open, loose groupings of dozens to a few thousand stars. Some open clusters, such as the Pleiades cluster in the constellation Taurus, are quite compact. Others are much looser and more difficult to distinguish from the surrounding random stars. Very loose, sparse groups of related stars are called associations. Open clusters tend to be young and sometimes contain much interstellar gas and dust. In some young associations, new stars are still being formed within noticeable clouds of gas and dust. *Globular clusters* are roughly spheroidal groups of up to millions of stars that formed at the beginning of their galaxy. Because globular clusters are so old, their brightest stars are red giants and supergiants. They contain little or no interstellar material because most of the primordial gas was used up in making stars. The Milky Way's globular clusters are usually outside of the galactic disk, in the surrounding area called the halo.

Nebulae   Nebulae (Latin for "clouds") are clouds of gas and dust in space, some relatively small in volume, others spreading

over hundreds or thousands of light-years. The gas consists mostly of hydrogen, with noticeable amounts of other elements, mostly helium, and even traces of water, formaldehyde, alcohol, and other compounds. The dust has carbon and silicon as its major components. Nebulae often have strange shapes, and many are named for things they (more or less) resemble—such as the Crab Nebula and the Horsehead Nebula. Bright nebulae are often beautiful sights to view through telescopes.

There are five main types of nebulae. *Dark nebulae,* or absorption nebulae, are dark clouds in which the dust absorbs or obscures the light of more distant stars. *Reflection nebulae* are illuminated by the glow of a nearby star. Since dust scatters blue light more effectively than red, a reflection nebula appears bluer than the star that is illuminating it. *Emission nebulae* shine brightly by fluorescence, which is caused by irradiation from very hot nearby stars. *Planetary nebulae* are thin shells of gas thrown off the surface of a star at the end of its red giant stage. *Supernova remnants* (SNRs), huge irregular shells of gas, remain when a supergiant star is destroyed in a supernova explosion.

Key to the Color Plates

Optical Aids

The following symbols, which appear with the text descriptions of the deep-sky objects, identify the minimum optical aid needed to see the object featured. They do not necessarily reflect the equipment used in making the photograph. Underneath the symbols for those objects accessible to amateur observers (with the naked eye, binoculars, or small to moderate-size telescopes), we have given the magnitude of the object, expressed as "m" followed by a number.

naked eye

binoculars

amateur telescope

observatory equipment

space telescope or satellite

The Deep-Sky Objects These illustrations appear on the following text pages to illustrate the type of object featured. Beneath each drawing we give the object's distance from Earth in light-years (ly); a question mark follows those distances that are uncertain.

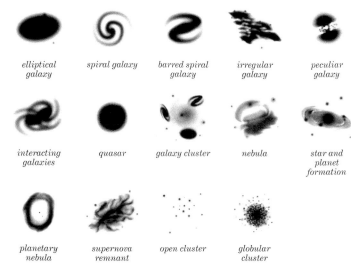

elliptical galaxy

spiral galaxy

barred spiral galaxy

irregular galaxy

peculiar galaxy

interacting galaxies

quasar

galaxy cluster

nebula

star and planet formation

planetary nebula

supernova remnant

open cluster

globular cluster

THE DEEP-SKY OBJECTS

m 8.0

60,000,000 ly

M49 (NGC 4472)
Elliptical galaxy, type E4, in Virgo

Elliptical galaxies, classified by their degree of flattening,
range from spherical, with no flattening (type E0), to very
flattened, almost disk-shaped (type E7). The classification
number is determined by measurements of the galaxy's
long and short axes. One of several giant elliptical
galaxies in the Virgo galaxy cluster, M49, type E4,
displays an intermediate degree of flattening. Its
diameter, approximately 60,000 ly, is about three-fifths
the diameter of the disk of our own Milky Way galaxy.
Like most ellipticals, M49 shows a notable absence of
interstellar gas and dust, necessary for star formation,
and a consequent lack of young, blue stars. Elliptical
galaxies, usually made up of older, cooler stars, are
yellowish in color, although this hue cannot be perceived
by the human eye at low light levels. Lacking the
exquisite structure seen in spirals, ellipticals appear to
be featureless groupings of stars, and many may lack
even a nucleus. Their stars move not in the choreographed
rotation seen in spirals, but at random, in different
directions and at varying speeds.

m 8.2

M32 (NGC 221)
Elliptical galaxy, type E3, in Andromeda

2,500,000 ly

M32, one of two major companions to M31, the Andromeda Galaxy, is a member of the Local Group, the weak cluster of perhaps 30 galaxies to which our Milky Way galaxy belongs. With a diameter of 5,000 ly, M32 is relatively small, but it is not as small as the dwarf ellipticals (dim, thinly populated galaxies that predominate in the Local Group). High-resolution images of its central core, taken by the Hubble Space Telescope in 1991, showed a high concentration of stars toward the very center—in the inner 100 ly the stars are more densely distributed than in our Sun's neighborhood by a factor of 100 million. This density is evidence that a supermassive black hole, of about 3 million solar masses (3 million times the mass of our Sun), resides at the galactic center. Despite its small size and its distance of 2.5 million ly, M32 has a magnitude of 8.2 and is visible in amateur instruments. In photographs of the Andromeda Galaxy, M32 is the small round galaxy closer to the giant spiral. M32's distance from M31 is unknown, but it may be in orbit around the larger galaxy, with an orbital period of perhaps 1 billion years.

24

Leo I
Elliptical galaxy, type E4, in Leo

600,000 ly

Dwarf ellipticals are the most common form of galaxy. Because of their relatively low brightness—they contain millions of stars as opposed to billions, as are found in typical spiral galaxies and some giant ellipticals—they can be hard to detect. Some are so thinly populated with stars that we can see right through them. Leo I is just 600,000 ly from our solar system, but its small size and low star count give it an apparent magnitude of only 10. It is part of the Local Group, the galaxy cluster to which our Milky Way belongs, as is its smaller dwarf elliptical companion, Leo II. Its low surface brightness and its proximity to the brilliant 1st-magnitude star Regulus (alpha [α] Leonis, only 20′ away), make Leo I difficult to see and photograph. The image shown here is the first color photograph of a dwarf elliptical galaxy. Color imaging, which reveals star colors—and, indirectly, their types and ages—can help astronomers understand star formation in this type of galaxy.

26

m 8.5

M87 (NGC 4486)
Elliptical galaxy, type E0, in Virgo

50,000,000 ly

One of three large elliptical galaxies near the center of the Virgo galaxy cluster, M87 is noted for its huge mass (about 40 times that of our Milky Way) and for the intriguing jet of plasma (hot, ionized gas) shooting outward from its core (shown in detail in the following plate). What appear to be stars swarming around the galaxy are some 3,000 globular clusters. The strongest radio emitter in Virgo, and one of the brightest in the sky, M87 has the radio designation Virgo A. Recent data gathered from the Hubble Space Telescope provide strong evidence of a black hole of 2 to 3 billion solar masses (1 solar mass is a unit equal to the Sun's mass) at M87's center. The relatively rare giant ellipticals such as M87 probably achieved their enormous masses by devouring smaller galaxies—galactic cannibalism. They tend to be found at the centers of rich clusters, with smaller surrounding galaxies offering plenty of prey. Some 50 million ly from our solar system, M87 covers a region of sky about 3' by 3'. Visible in 3- to 8-inch telescopes, it is better appreciated with larger scopes.

M87 (NGC 4486)
Close-up of core and inner jet

50,000,000 ly

The enormous source of power of the energetic elliptical galaxy M87 (preceding plate) is to be found at its heart. With the improved optics of the Hubble Space Telescope, in 1994 observers determined that material surrounding the very center is moving at too high a speed to remain there without the gravitational pull of a supermassive black hole. Ultraviolet images (bottom) provided evidence of high-speed electrons, indicating that something had blasted the atoms to high speeds. The strong light source at the very center of M87 can be seen in the orange-tinted infrared image (top). (The scattered dots of light are globular clusters.) The 5,000-ly-long jet of plasma, displayed in both images, and the prodigious quantities of electromagnetic energy outpouring from the center indicate that M87 was involved in an enormous explosion, perhaps a collision with a clump of hot gases or another galaxy. The core also emits intense quantities of radio energy. There may be a twin jet on the other side of the galaxy, and M87 may be a developing two-lobe radio source, such as Centaurus A and NGC 4261.

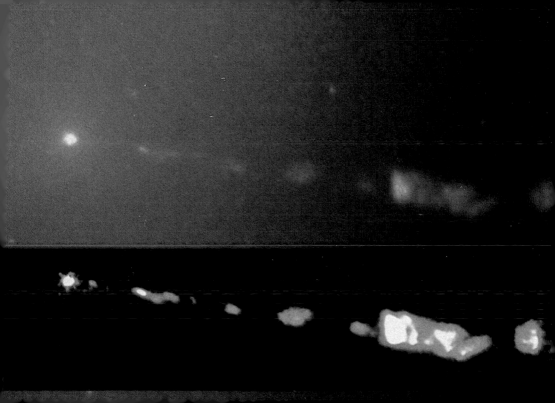

NGC 4261 (core and jets)
Elliptical galaxy, type E3, in Virgo

45,000,000 ly

A giant elliptical galaxy in the Virgo Cluster, NGC 4261 displays two radio jets, giant lobes of radio energy emitting from the galaxy in opposite directions (photo on left). Such lobes result from an explosive event or from continually infalling material at the galaxy's center being accelerated and ejected by a central black hole. Astronomers have hypothesized that a thick, doughnut-shaped disk of gas and dust surrounding the center, called the accretion disk, "feeds" the black hole, a region of such enormous density that nothing, not even light, can escape its intense gravitational pull. The image on the right, of the galactic center, shows the accretion disk to be about 300 ly across. New evidence has indicated that supermassive black holes on the order of millions of solar masses may exist at the center of many, perhaps most, galaxies. These two Hubble Space Telescope photographs support the theoretical picture of galactic black holes that scientists have developed over the past decade, such as the perpendicular orientation of the radio jets with respect to the disk of the galaxy's central region.

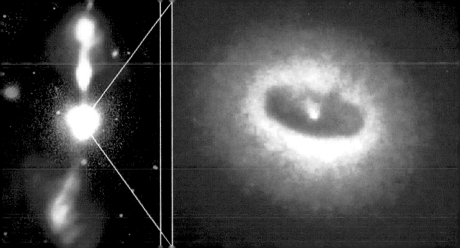

m 9.9

NGC 4753
Spiral galaxy, type S0 peculiar, in Virgo

50,000,000 ly

NGC 4753 represents a class of galaxies that share
properties of both ellipticals and spirals and perhaps
represent an intermediate or transitional phase between
the two. Called type S0 (or lenticular) galaxies, they exhibit
the central bulge and flat disk of spirals, but lack spiral arms
and interstellar gas and dust, as do ellipticals. The galaxy
classifier Edwin Hubble placed S0 galaxies, with their
apparently hybrid nature, at the juncture between
ellipticals and spirals in his diagram of galaxy morphological
types. The concentrated gases and dust, young blue stars,
and pink hydrogen-emission clouds—indicative of star
formation in spiral galaxies with arms—are lacking in
type S0 galaxies, as they are in ellipticals. These galaxies
are generally found within dense, rich clusters that
display a high proportion of elliptical and S0 galaxies
to normal spirals. Some astronomers believe S0 galaxies
result when a galaxy passes through a spiral galaxy,
drawing out dust and gas but leaving the stars intact.
NGC 4753 is classified peculiar because of the atypical
bands of dust, an indication that a collision is occurring.

34

m 8.3

M104 (NGC 4594) Sombrero Galaxy
Spiral galaxy, type Sa, in Virgo

Spiral galaxies are classified by the size of their nuclei—central bulges composed of a high concentration of old stars and some fast-moving gases—relative to the size of the surrounding disk, and the tightness of their spiral arms. Types range from Sa, with relatively large, bright, dominant nuclei and tightly wound arms, to Sc, with weak nuclei and large, bright, loosely wound spiral arms. The Sombrero Galaxy is the nearest (about 30 million ly from Earth) and brightest (8th magnitude) of the Sa galaxies. Because of its orientation, we view this galaxy almost edge-on; dust clouds in the plane of the galaxy's disk block our view deeper into the galaxy. Spiral galaxies have relatively little interstellar gas and dust in their central bulges, which are populated mainly by older stars. Their disks, however, are usually rich in gas and dust, providing the conditions necessary for star formation. It is in the disk of our Milky Way that many of the nebulae and star clusters familiar to stargazers lie, and astronomers surmise that the disks of other spirals feature similar objects.

30,000,000 ly

M31 (NGC 224) Andromeda Galaxy
Spiral galaxy, type Sb, in Andromeda

The great Andromeda Galaxy, 2.5 million ly from our solar system, is the most distant object visible to the naked eye. This spiral galaxy, with an estimated half-trillion or more stars, is the largest member of our Local Group, with perhaps twice the mass of the Milky Way galaxy. Astronomers think our Milky Way, a type Sb or Sbc, looks similar to M31. Both have a fairly large central bulge surrounded by an extensive disk and spiral arms rich in interstellar gas and dust, where new stars form. The two galaxies are heading toward one another in their orbits, moving about 4,800 km (3,000 miles) closer every minute—a snail's pace in intergalactic terms. The Andromeda Galaxy has two notable companion galaxies (both visible in the photograph), M32 (NGC 221), a 9th-magnitude elliptical galaxy right beside it, and M110 (NGC 205), a dwarf, more flattened elliptical galaxy farther out. M31 appears to the unaided eye on clear, dark nights as a faint oval of fuzzy light (because of its hazy appearance, it was long considered a nebula). A telescope is needed to see its companions.

2,500,000 ly

M31 (close-up of double nucleus)
Spiral galaxy, type Sb, in Andromeda

When astronomers trained the Hubble Space Telescope on the nucleus of the Andromeda Galaxy in 1991 they found quite a surprise: not one, but two centers. The brighter one, visible from Earth, turns out to be not quite at the very center of the galaxy. At that location is another, dimmer concentration of stars. The distance between the two is only 5 ly (about the distance of our Sun from its nearest star). While it is possible that there is really only one nuclear center with the middle dimmed by thick dust clouds, astronomers think it likelier that the two concentrations result from a rather recent collision of the Andromeda Galaxy with another galaxy that has left the cores of both remaining in the central regions. This has caused scientists to theorize that if an apparently ordinary galaxy like M31 underwent such a major event, galaxy collisions may be much more common than had been realized. Meanwhile, they are studying the nucleus of M31 for conclusive evidence of a supermassive black hole there, comparable to one that evidence hints is at the center of the Milky Way.

2,500,000 ly

40

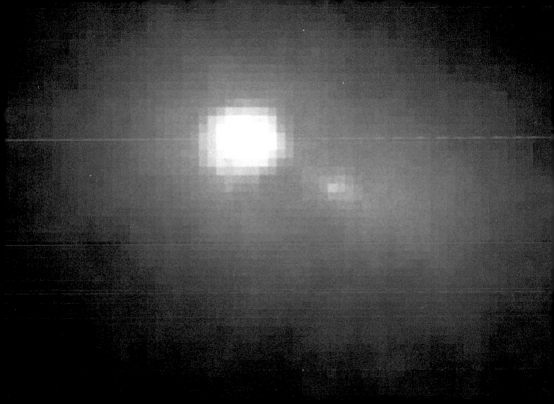

NGC 4565
Spiral galaxy, type Sb, in Coma Berenices

Sometimes called the "flying saucer" galaxy, NGC 4565 offers us a distinctive, almost perfectly edge-on view. A type Sb galaxy with a 90,000-ly diameter, it is comparable in size to the Milky Way and probably shows us how our galaxy would look from the same perspective. Spiral galaxies are notable for the dust clouds concentrated in the planes of their disks, apparent as a dark lane of material in NGC 4565. Numerous forces—gravity, magnetic fields, stellar winds, collisions, and a shearing effect caused by their revolution around the galaxy's center—act on these clouds, giving them their often ragged appearances. Edge-on galaxies such as NGC 4565 also enable astronomers to measure the radial velocity, the motion toward us or away from us along our line of sight, of the stars in their disks; the more edge-on a galaxy is oriented, the more easily radial velocity can be measured. Such measurements led astronomers to hypothesize the existence of undetectable matter, dubbed dark matter, around galaxies, because the visible matter in the universe cannot account for certain gravitational effects observed.

20,000,000 ly

M100 (NGC 4321)
Spiral galaxy, type Sc, in Coma Berenices

50,000,000 ly

The central region of the nucleus of this Virgo Cluster member (the Virgo Cluster extends into the constellation Coma Berenices) is shown in this Hubble Space Telescope image. M100 shows a concentration of hot, young stars in this area, where older, yellow stars are usually found. The nucleus is small compared to the enormous area covered by the outer disk and arms, a characteristic of type Sc spiral galaxies. The widespread bluish glow and the occasional pinkish hydrogen emission clouds seen here indicate considerable star-formation activity. Stars form from gas and dust clouds, which are concentrated in the arms of spiral galaxies. Stars of all types form, and although cool, red, low-mass stars form in greater abundance, the hot, blue, high-mass stars shine more brightly, giving spiral arms their bluish hue. The pinkish splotches are regions that are especially thick with hydrogen clouds from which a number of high-mass stars have just formed; the resulting ultraviolet radiation from these stars causes the hydrogen to fluoresce in the visible, mostly red, region of the light spectrum.

44

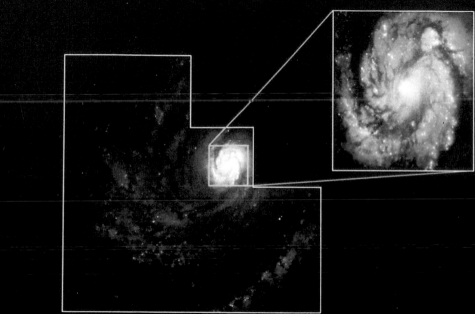

m 5.7

M33 (NGC 598)
Spiral galaxy, type Sc, in Triangulum

Known as the Pinwheel Galaxy, M33 has the weak nucleus and extensive arms typical of Sc spirals. It is the third-brightest member of our galaxy cluster, the Local Group, outshone by the Andromeda Galaxy (M31) and our own Milky Way, both type Sb spirals. (Most of the rest of the Local Group galaxies are dwarf ellipticals.) About as far from us as the Andromeda Galaxy (2.5 million ly), M33 is fainter, with roughly one-fortieth the number of stars. Its disk is about 50,000 ly across, compared with the Andromeda's diameter of 150,000 ly, and the Milky Way's 100,000 ly. Astronomers suspect that the long-term survival of spiral galaxies depends upon whether they are members of richly populated galaxy clusters, where they can interact and even collide, which results in galactic mergers that make giant and supergiant ellipticals. The Local Group is a sparse cluster with no giant ellipticals, and the odds of major collisions are low, so we can expect these three spirals to survive for some time—yet new evidence suggests that the Andromeda Galaxy may have collided with a dwarf galaxy not long ago.

2,500,000 ly

m 8.4

M51 (NGC 5194), Whirlpool Galaxy, and **NGC 5195**
*Spiral galaxy, type Sc, and irregular galaxy
in Canes Venatici*

M51, the Whirlpool Galaxy, lies at the center of a small group of galaxies in the constellation Canes Venatici and was the first clearly recognized spiral galaxy. The smaller, irregular companion galaxy (possibly once a spiral itself), NGC 5195, clearly has interacted with M51, pulling its outermost arms askew. Although it looks as if one of M51's spiral arms connects the two, the arm is actually in front of NGC 5195 (from our perspective), far out of the companion's reach. The two galaxies probably swept past one another millions of years ago, each distorted by the gravitational forces of the other, with the smaller getting the worst of it and losing any semblance of an orderly shape. M51 has many young blue stars in the arm closest to NGC 5195. Star formation may have been spurred on by the galactic interaction. This recent image of the two galaxies was compiled from three different color recordings. The resulting images were then combined to produce a color image approximating what the human eye would see when looking at the pair, given sufficient light to perceive color.

40,000,000 ly

48

m 7.0

NGC 253
Spiral galaxy, type Sc, in Sculptor

13,000,000 ly

A spiral galaxy with the large, well-developed disk and the small, weak nucleus of type Sc galaxies, NGC 253 lies at an oblique angle of only about 10° to our line of sight. It is more than 13 million ly away in the Sculptor Group, the closest galaxy cluster to our cluster, the Local Group. Astronomers have noted more dust here than is usual for spiral galaxies, but have yet to come up with an explanation for it. This dust, which is produced as generations of stars fuse hydrogen, helium, and heavier elements and blow the processed material back to space, is more noticeable toward the inner regions of the disk, where it thereby obscures details of the galactic center. The outpouring of visible light from the nucleus is dimmed significantly by the absorbing dust and is reradiated in prodigious amounts at lower temperatures as infrared radiation. As in all Sc galaxies, the spiral arms of NGC 253 appear blue, an indication of large-scale star formation. Astronomers are trying to determine whether spiral galaxies evolve, perhaps starting out as type Sa and evolving through Sb to Sc.

50

m 7.6

M83 (NGC 5236)
Barred(?) spiral galaxy, type SBc or Sc, in Hydra

Astronomers disagree on whether M83 displays the
central bulge of a standard spiral galaxy or of a barred
spiral. They do agree that this is an unusually active
galaxy, currently especially busy with the formation of
stars. In this beautiful image the spiral arms display the
characteristic blue color that indicates an abundance of
young, hot, blue stars, which outshine all others, along
with the accompanying red clouds of excited hydrogen
gas, both features associated with star formation. Another
indication of star-forming activity is that supernovas,
which are the explosive deaths of the very short-lived,
most massive stars, are observed in M83 more frequently
than in any other known galaxy. The dynamics, forces, and
motions of this galaxy are unusual. Some astronomers
have suggested that they may have been spurred by a past
close encounter with another galaxy that also warped the
spiral arms with respect to clouds of cold hydrogen gas
(shown in radio maps) surrounding the galaxy. Such an
encounter might be expected to instigate large-scale star
formation in colliding gas and dust clouds.

2,200,000 ly

NGC 1365
Barred spiral galaxy, type SBb, in Fornax

100,000,000 ly

Spiral galaxies offer us beauty in space. In barred spirals, stars in the central bulge form an elongated bar that extends to the rim of the disk. This unusual arrangement makes barred spirals both visually and intellectually stimulating, as astronomers try to solve the mysteries of their large-scale symmetries. One theory is that the bar is a temporary phase, occurring when the rotating stars in a spiral galaxy "pile up" in a line across the diameter of the disk. Barred spirals are classified in the same way as spirals, by the size of the central bulge relative to the disk and by the tightness of the spiral arms. NGC 1365 is the largest spiral galaxy in the Fornax Cluster, which is found in the direction of the southern constellation Fornax (the Furnace). It is about as massive as our Milky Way, itself a substantial galaxy. The colors in the photograph show the predominance of older, yellowish stars in the nucleus and bar; the young, blue stars and glowing pink hydrogen-rich nebulae associated with star-forming regions are evident in the spiral arms.

NGC 5383
Barred spiral galaxy, type SBb, in Canes Venatici

75,000,000 ly

Even in the black-and-white images astronomers use to bring out finer detail, barred spirals are fascinating to behold. This image of NGC 5383 reveals the complicated absorption patterns brought about by the substantial dust lanes lying in the bar. About one in five spiral galaxies is a barred spiral, but astronomers are still not sure how or why the bar comes about. It may be a temporary phase through which some or all spirals go, resulting from a temporary, random, large-scale gravitational instability that can be expected to arise from time to time in a large rotating mass of stars. The bar rotates with the rest of the galaxy and tends to be similar to the central bulge of a standard galaxy, composed mainly of older yellow stars. Because our time in the universe is infinitesimally brief, we are unable to witness the unfolding of events that may precede the formation of a barred spiral galaxy or any other type. It may take millions of years for a bar to form, so we essentially see such galaxies in a freeze-frame. Barred spirals are elegant reminders of how much there is yet to learn about our universe.

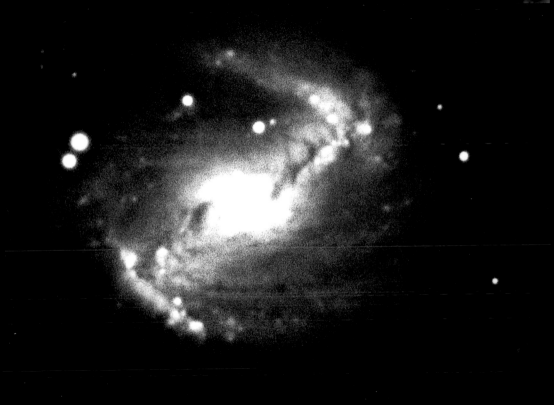

NGC 1566
Spiral galaxy, type Seyfert, in Dorado

50,000,000 ly

Seyfert galaxies are spiral galaxies with an extremely bright nucleus that emits high amounts of energy across the spectrum. NGC 1566 is the brightest member of a galaxy cluster only 50 million ly from our solar system. Although the galaxy's central bulge appears relatively large, that apparent size is a function of brightness—much of the light from the nucleus comes from an incredibly small region, possibly from a supermassive black hole on the order of millions or billions of solar masses. Theory predicts that matter falling into the black hole would experience an explosion, blasting out energy all across the spectrum, as seen in Seyferts. Having made advances in the study of quasars, astronomers have concluded that Seyfert galaxies are intermediate between normal spiral galaxies and quasars. Seyferts evidently have less infalling matter than quasars but more than normal spirals. This matter could have come from a (cosmically) recent collision with a dwarf galaxy.

NGC 6822
Irregular galaxy, type Irr I, in Sagittarius

1,600,000 ly

This dwarf irregular galaxy is the most isolated member of our Local Group. At 1.6 million ly from Earth, it is close enough that its brightest stars can be resolved in a moderately large telescope. Its largest dimension is about 11,000 ly across, a factor of ten smaller than the Milky Way. Galaxies classed as irregular are those that show no clear symmetry. They are grouped in two categories, based on observational differences (dependent upon distance) and real differences. Type Irr I galaxies are close enough to enable us to resolve individual type-O and type-B stars, as well as emission nebulae, which are indicative of ongoing star formation. The best-known examples of this type of galaxy are the Large and Small Magellanic Clouds. Type Irr II galaxies exhibit no standout stars or glowing nebulae and have prominent dust lanes. M82 and NGC 5195, the companion of M51, are examples of type Irr II. Many irregular galaxies are satellites of larger galaxies, and in many cases it is thought that the gravitational attraction of the larger galaxy has pulled them into their odd shapes.

m 1.0

Large Magellanic Cloud
Irregular galaxy, type Irr I, in Dorado

170,000 ly

When the survivors of Ferdinand Magellan's round-the-world voyage returned to Europe in 1521, they reported sighting two ghostly clouds in the night skies of the Southern Hemisphere. Around 1930 astronomers discovered that these "clouds" were galaxies. Not only are they members of our Local Group, these type I irregulars are satellite galaxies of our Milky Way, gravitationally bound and orbiting around it. Some astronomers speculate that the Magellanic Clouds were at one time barred spirals—until pulled asunder by our own galaxy. The Large Magellanic Cloud (LMC), 170,000 ly from Earth, extends 60,000 ly at its widest point and contains perhaps 15 billion stars. A hydrogen bridge connects the LMC with the outer disk of the Milky Way, and a huge loop of hydrogen gas, called the Magellanic Stream, surrounds both Magellanic Clouds. These factors suggest that the LMC has passed through the Milky Way's disk. Star formation is going on at a high rate in the LMC, possibly as a result of that passage.

62

m 2.5

Small Magellanic Cloud
Irregular galaxy, type Irr I, in Tucana

200,000 ly

Another dwarf type Irr I companion galaxy to the Milky Way, the Small Magellanic Cloud (SMC), lies about 23° away from the Large Magellanic Cloud (LMC), and only 30,000 ly farther out (200,000 ly compared to the LMC's 170,000 ly). The Magellanic Clouds appear to be related, perhaps split apart in some previous passage through the disk of the Milky Way. They are immersed in a common hydrogen cloud (the Magellanic Stream) that trails far away from the SMC. As the smaller of the two galaxies (one-eighth the diameter of the Milky Way, compared to the LMC's one-sixth), the SMC is also fainter, but both are easy to see from southern latitudes. Recent radio and optical studies indicate that the SMC is split into two sections along our line of sight, with the two subclouds separated by about 20,000 ly. Like the LMC, the SMC is characterized by active star formation and exhibits resolvable bright stars.

m 8.4

M82 (NGC 3034)
Irregular galaxy, type Irr II, in Ursa Major

A type II irregular galaxy, M82 displays no resolvable stars and is heavily laden with dust lanes. It is also classified as peculiar because of high-energy activity detected behind its curtain of dust. Strong radio emissions are pouring out of its nucleus, and much excited (energized) gas is streaming out of the central regions, which suggest that there has been an enormous explosive event in the nucleus of this galaxy, perhaps a gas cloud from a dwarf companion galaxy colliding with a central supermassive black hole. M82 has an overall spectral class of type A, but its individual stars are too faint to be seen. It appears to lack the young, blue type-O stars, the hotter type-B stars, red supergiants, and brighter giants that are associated with active star formation. It could be that star forming simply hasn't begun or become apparent yet, or that it is hidden behind the thick clouds of dust. Some astronomers think M82 is a spiral galaxy presented to us in profile that has been distorted by the gravitational field of the large spiral M81, only 100,000 ly away.

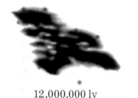

12,000,000 ly

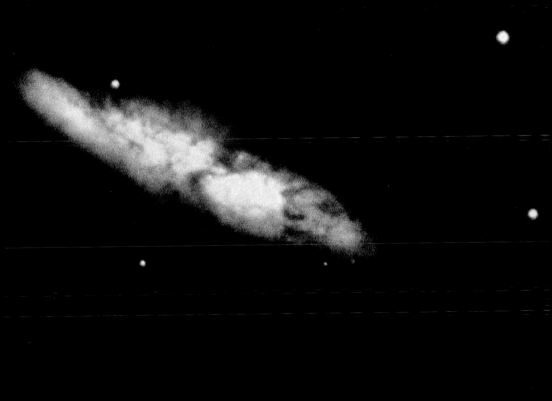

m 7.0

NGC 5128 Centaurus A
Peculiar galaxy, type E0, in Centaurus

13,000,000 ly

Centaurus A is one of the most powerful emitters of radio energy in the sky (the "A," a radio designation, labels it as the most powerful radio emitter in Centaurus), a substantial source of X rays and gamma rays, and one of the most luminous and massive galaxies known. Only 13 million ly from Earth, it is the most active, powerful galaxy in the Local Group's vicinity, but great dark dust lanes obscure it from the naked eye. The exciting picture that is emerging from studies of Centaurus A is of a supergiant spheroidal elliptical galaxy (E0) cannibalizing a spiral galaxy, whose outer rim we observe as the dust band across the elliptical's middle. There is also evidence of a supermassive black hole at the center, suggested by infalling material blasting out from the center in two enormous radio lobes well outside and at opposite directions from the visible galaxy. The nucleus is variable, with energy outputs varying over brief intervals, an indication that the galactic center occupies a very small space, further evidence that it is powered by a supermassive black hole.

Ring Galaxy
Peculiar galaxy in Volans

Ring galaxies are rare peculiar galaxies named for their circular patterns. The ring is the temporary result of the collision of two galaxies, most likely a small galaxy passing through the center of a spiral. Once the intruder has departed, the stars and other material left behind in the spiral, liberated from its gravitational pull, fly outward, pushing the outer material into the ring form. It has been hypothesized that eventually such galaxies will reorganize into a spiral pattern. Ring galaxies represent only a small fraction of the galaxies that show evidence of a collision because they result from a rare occurrence— a centered, frontal collision. Given galaxies' velocities and the ratio of their sizes to intergalactic distances, galactic collisions of all types can be expected over long periods of time. The ratio of a star's size to typical interstellar distances is insignificant, however, and considering the space velocities of stars, it has been estimated that no two stars have ever collided in the universe, even during collisions of galaxies.

250,000,000 ly

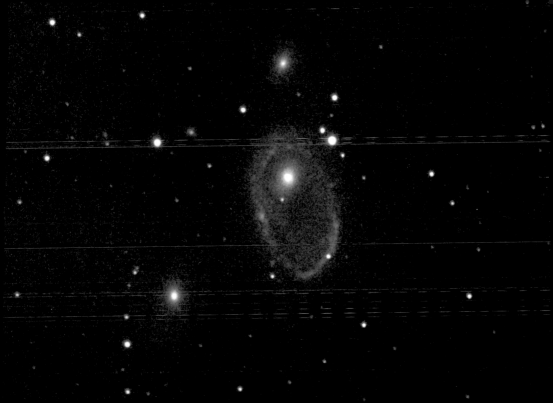

NGC 2685 Helix Galaxy
Peculiar galaxy, type S0, in Ursa Major

30,000,000 ly

NGC 2685 is another example of a peculiar galaxy that appears to be the intermingling of two galaxies, perhaps two spirals. Recent large-scale studies of galactic evolution indicate that over time the number of spiral galaxies is diminishing while the number of giant ellipticals is increasing. This is due either to some internal evolution of spirals or to collisions of spirals with giant ellipticals or other spirals. Such collisions could certainly destroy the clear, gravitationally induced symmetry of a spiral galaxy and affect its angular momentum, resulting in a giant elliptical form. It is tempting to speculate that we might be observing just such a process in the Helix Galaxy, in which two axes of symmetry are apparent, as is an encircling ring. We can only speculate what will come of this coupling, say 100 million years from now. It is interesting to note that although gravitational interactions abound in a collision of two galaxies, and although the vast interstellar nebulae within each galaxy will ram into clouds of the other galaxy, the individual stars are far too small relative to interstellar distances to ever hit one another.

NGC 4676a and b "Mice" Galaxies
Peculiar galaxies in Coma Berenices

200,000,000 ly

This pair is one of the best known of the colliding galaxies. Their nearness to one another, the bridge that appears to connect them, and their tails are the evidence that led scientists to conclude that they are in the act of colliding. Discoveries of such interacting galaxies opened our eyes to a universe full of activity. Astronomers began to see that galaxies come together, often passing each other with distortions to both and sometimes colliding. Computer studies showed that with the proper selection of approach velocities, researchers could generate images of galactic interactions that look virtually identical to what they see through telescopes, thus confirming theoretical predictions. In this image of NGC 4676a and b, false colors are used to depict different levels of brightness; yellow signifies lowest brightness levels, while light blue, red, and dark blue range at the bright end. Details of this interaction are thereby brought out better, including the tails, which form when tidal interactions (unbalanced gravity forces) brought on by the collision cause the ejection of material (stars and nebulae) from the galaxies.

74

NGC 4038/39 "Antennae" Galaxies
Peculiar galaxies in Corvus

50,000,000 ly

NGC 4038 and 4039, another fascinating pair of galaxies caught in the act of colliding, are clearly two merging, yet still distinct galaxies. Both galaxies appear to be spirals, although one is of the gasless, armless S0 type. The two long tails of material (only one is faintly visible), called "antennae," are stars and interstellar gas and dust that were left behind when the two galaxies met and each lost its gravitational hold on its components. (Interacting galaxies with long tails such as these are also called "rattail" galaxies.) The tails might extend half a million light-years from tip to tip. The true color of this image of NGC 4038/39 brings out interesting aspects of each galaxy. One disk is bluer, typical of a normal galaxy with healthy star formation in its spiral arms. The other, on the right, is the S0, appearing somewhat yellowish, indicative of an older star population in which recent star formation is lacking. The increase in star formation brought on by the collision is shown by the pink hydrogen emission clouds and clouds of dust. The two yellowish nuclei, containing mainly older stars, are still distinct.

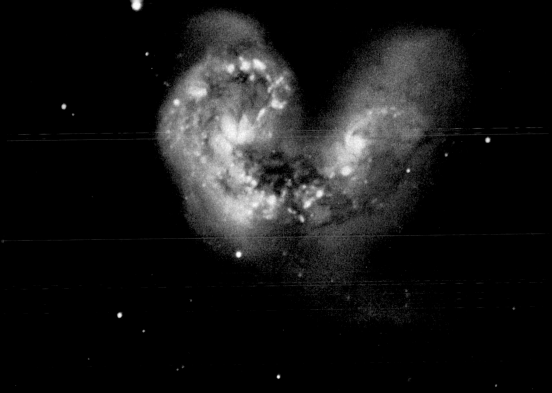

ESO B138 + IG 29/30 "Toadstool" Galaxies
Peculiar galaxies in Ara

The "Toadstool" galaxies are located in (actually beyond) the southern Milky Way, in the small constellation of Ara, the Altar. It is rare that we can see galaxies and other deep-sky objects along the Milky Way, because of the presence there of a considerable amount of light-absorbing dust; this area happens to be relatively clear. The sky near the plane of the Milky Way has been called "the zone of avoidance," because it looks as if other galaxies avoid it. In this image, details are brought out by the false-color technique. A strong bridge of hydrogen gas streams between these two galaxies, suggesting they have already collided and are now separating. New star formation as a result of this close encounter has been detected in this bridge. One might wonder how such a collision would affect planets in the two galaxies. Planets near areas where nebulae collide might suffer from the resulting star-forming activity, supernova explosions, and stellar winds. Others will by and large remain unaffected, as will most individual stars.

190,000,000 ly

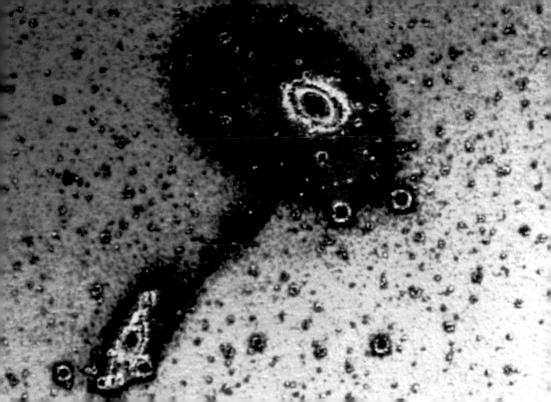

G2237 + 0305 Einstein Cross
Gravitational lens and quasar in Pegasus

These five "objects," actually the light from two sources, illustrate the part of the theory of relativity that states that light passing by a strong gravity source will be bent. What we see here is the light of a single quasar, 8 billion ly away, that has been bent in multiple ways as it passed through an intervening galaxy (400 million ly away), which acts as a "gravitational lens." At the center is the nucleus of the lensing galaxy; the four bright outer lights are the multiple images of the quasar. Found billions of light-years out in space at the maximum distances observable, quasars may arise sporadically from the collision and merging of galaxies, but the farthest (and oldest) were formed 1 to 2 billion years after the universe's origin, when galaxies were taking shape. Researchers have concluded that quasars contain supermassive black holes, formed when huge amounts of infalling hydrogen and helium crashed together at the centers of fledgling galaxies, compressing the centers to incredible density and blasting out matter and energy like an ongoing super-supernova.

8,000,000,000 ly

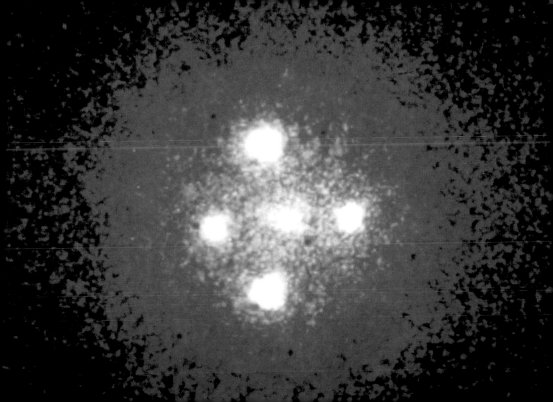

NGC 1275
Peculiar galaxy in Perseus

300,000,000 ly

NGC 1275 is a messy giant elliptical galaxy pouring out
enormous amounts of radio energy plus about one-fifth of
the X rays emitted by the entire Perseus galaxy cluster.
The ground-based telescope photograph on the left, as
well as the galaxy's radio output, led astronomers to
suspect that NGC 1275 is undergoing the aftereffects of
a collision. The high-resolution Hubble Space Telescope
image on the right provides evidence to support that
theory. The white spot at the center is the compact high-
energy core, presumably a supermassive black hole. The
blue dots are globular star clusters. All globular clusters
known in the Milky Way and around other galaxies are
yellowish because of their extreme age. The blue color of
these globulars implies that they are young or recently
have experienced the formation of hot, young, blue stars.
In either case, they provide evidence of major, widespread
star-forming activity, as do the many light-absorbing dust
clouds in this galaxy.

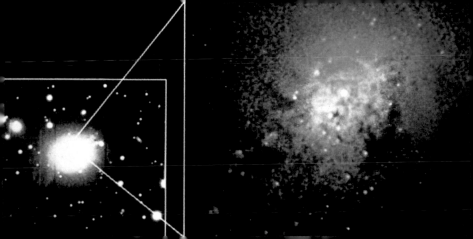

3C 273
Quasar in Virgo

2–3,000,000,000 ly

3C 273 is one of the brightest quasars in our sky, although it is hundreds of times too faint for the naked eye to perceive. It was the first quasar to be studied optically, in the early 1960s, after radio astronomers had noticed its radio output (only 3 percent of quasars are powerful radio emitters). In visible light it puts out several hundred times more light than the Milky Way, yet most of this light comes from a volume of space only about as large as our solar system. A shock-absorbing, rotating torus (or doughnut) of gas and dust surrounds the galaxy's core, and ejected material moves out most successfully along the rotational poles of the torus. 3C 273, curiously, shows only one jet in this visible-light image. The other jet is perhaps buried in nebulosity, a feature of active galaxies. The presence of jets of material blasting out from the core of a galaxy indicate that an explosive event has recently occurred there. As is evident in this photograph, the quasar (center) looks much like the ordinary stars around it. It is because they appear as faint stars that quasars (also called "quasi-stellar objects") long eluded the attention of astronomers.

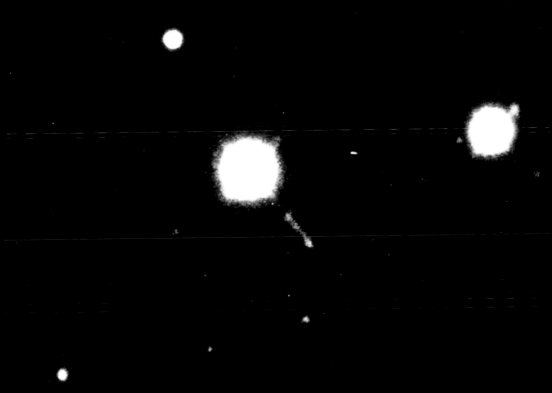

3C 275.1
Quasar in Virgo

3C 275.1, a quasar at the center of a galaxy cluster, is accompanied here by cluster galaxies and foreground stars. It is the large object near the center flanked at 7 o'clock and 2 o'clock by smaller objects, probably cluster galaxies. Its redshift (the shifting of spectral lines toward the red end of the spectrum in very distant objects, used to measure an object's distance) places it roughly 7 billion ly away. The light that reaches Earth left the quasar 2 billion years before our solar system came into existence. This false-color image of 3C 275.1 reveals a small, glowing halo around the quasar. Upon realizing by the mid-1960s that quasars were very distant and luminous objects, astronomers were further challenged when evidence indicated that they were as small as our solar system. Astronomers were fascinated by these exceedingly small objects, which were even more powerful than Seyfert galaxies. The detection and examination of surrounding light eventually showed that these powerhouses were located at the centers of otherwise normal galaxies.

7,000,000,000 ly

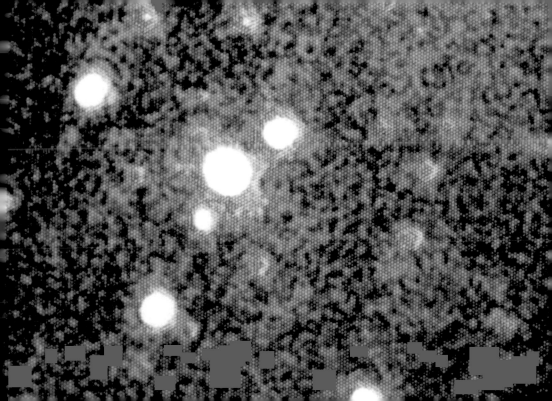

QSO 0351 + 026
Quasar-galaxy pair in Taurus

750,000,000 ly

Since the 1970s astronomers have hypothesized that the enormous redshifts of quasars were truly indicative of their great distances, and that they were extremely small objects (about the size of our solar system) that radiated hundreds of times more energy than normal large galaxies. In interpreting photographs that seemed to show quasar-galaxy pairs, if the redshifts of the two objects were different, astronomers assumed that the redshift (not the photograph) told the true distance, and that the two were indeed far apart, only appearing related from our line of sight. This false-color image shows a galaxy and quasar clearly connected; their redshifts are the same and support this conclusion. The quasar (on the left) is possibly an example of a quasar formed by a galactic collision. The interaction may have dumped considerable material into the central regions of the quasar's precursor, wherein lay a supermassive black hole. The resulting acceleration of material around the center may have led to violent explosions that light up the central region as a quasar.

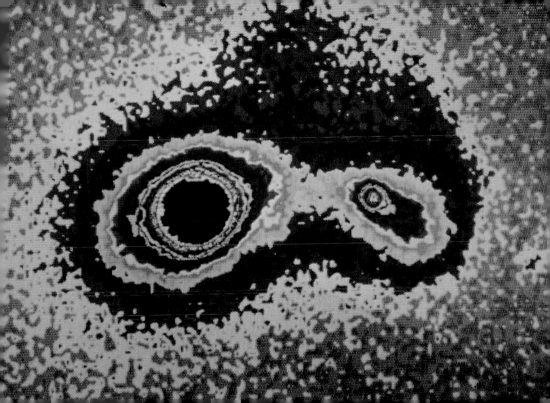

Virgo Cluster
Galaxy cluster in Virgo

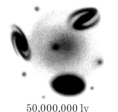

50,000,000 ly

The Virgo Cluster, the nearest moderately rich galaxy cluster, is 50 million ly away, close enough to have a gravitational influence on us. It is thought to lie at or near the center of the supercluster of galaxies that includes the Local Group, the cluster to which our Milky Way belongs. Rich, spherical clusters like this one, with 3,000 galaxies (mostly dwarfs), usually have one or more giant ellipticals near their centers, which have apparently grown from collisions and mergers, a process called galactic cannibalism. (Poor, shapeless clusters like the Local Group, with some 30 members, have a higher proportion of spiral galaxies.) Three large ellipticals dominate this cluster—M86, near the center; M84, on the right; and the supergiant M87, not shown. The richness (population density) of this cluster increases the likelihood of galactic collisions. M87's supermassive black hole and plasma jet are thought to be evidence that the galaxy has suffered at least one collision. On the left side of this image we can see a galactic interaction in the distortion of NGC 4438 by NGC 4435.

Coma Berenices Cluster
Galaxy cluster in Coma Berenices

The Coma Berenices galaxy cluster is a rich, spherical cluster with about 3,000 members, predominately elliptical and type S0 spiral galaxies. At a distance of 300 million ly, dwarf galaxies, which comprise the majority of galaxies, are not visible; only the spirals and giant ellipticals can be seen (most of the objects visible here are galaxies, a few are stars). S0 and elliptical galaxies generally lack the dust and gases associated with active star formation and are made up of older stars, which give the galaxies their yellowish color. The galaxies in dense, rich clusters such as this one have more opportunities to interact. Although galactic interactions can stimulate star formation, they can also sweep out the gases necessary for star formation. This might explain how S0 galaxies, flat disks lacking interstellar gas and dust and spiral arms, come about. As the universe evolves, the beautiful armed spiral galaxies are decreasing in number, possibly because of such galactic interactions.

300,000,000 ly

Stephan's Quintet
Galaxy cluster in Pegasus

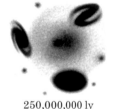

250,000,000 ly

This poor, irregular cluster of galaxies, discovered in 1877 by French astronomer M. E. Stephan, presumably contains dozens to hundreds of yet undetected dwarf galaxies. Of interest because of the clear interactions among several of the brighter galaxies, it became more intriguing when the redshifts (the wavelength expansion to longer waves, caused by the motions of galaxies away from us) for the five major galaxies were determined. Hubble's Law states that redshift increases with greater distance. Four of the galaxies, as would be expected, had the same redshift; one of those four is a large elliptical galaxy just beyond the right edge of the image. But the fifth had a lower redshift, indicating that it was ten times closer. A few astronomers have used Stephan's Quintet as evidence for questioning the reliability of redshifts as indicators of distance. But most have simply concluded that Stephan's Quintet is in reality a quartet, and that the closer galaxy, the blue spiral at the bottom, does look more detailed, less distorted, and somewhat larger than the others.

Cosmic Background Radiation
All-sky radio map

The beginning universe was very smoothly distributed in matter and energy, but astronomers weren't sure how it went from its smooth state (thermodynamic equilibrium) to the "clumpiness" that would have been necessary for galaxies to start forming within 1 to 2 billion years. Recently astronomers have begun to perceive enormous structures, such as galaxy clusters and superclusters, voids (huge "bubbles" lacking galaxies), and the vast "Great Wall" of galaxies. For these structures to have arranged themselves in less than 20 billion years—the estimated age of the universe—there had to have been small irregularities in the very early universe. The Cosmic Background Explorer was sent into space in 1989 to measure the faint cosmic background radiation left over from the Big Bang and to search for any irregularities. This all-sky radio map depicts the irregularities, with the red indicating slightly warmer and denser areas than the blue. A problem remains: The irregularities are much smaller than astronomers had expected based on current theory on the evolution of the universe.

Milky Way Galaxy
Mosaic of the galactic plane

This beautiful mosaic of the entire sky, mapped with respect to
the plane of the Milky Way, represents our view, absent the Sun,
from inside our type Sb spiral galaxy. We view it edge-on, and the
bright disk, splotched with light-absorbing dust clouds, envelops
us. The center of the Milky Way is toward the center of this
mosaic, the north galactic pole is at the top, and the south galactic
pole is at the bottom. The map wraps around, so that the extreme
left meets the extreme right. Although we cannot see all the way
to the galactic center, its brightness and bulge—in the direction of
the constellation Sagittarius—are evident. The bright region in the
plane to the left, midway to the edge, is in the constellation Cygnus.
It is part of our spiral arm and relatively near us. The Large and
Small Magellanic Clouds, our companion galaxies, are below the
galaxy's plane, right of center. The bright dot just beneath the
plane at the far right is the brightest star in our night sky, Sirius,
in Canis Major. Below and to the right of Sirius is Orion, slightly
squeezed by this projection.

Milky Way Galaxy
Infrared view of the galactic plane

Compare this false-color infrared image of the galactic plane with the visible-light view shown in the previous plate. Infrared light is emitted by warm regions, usually where there is ongoing star formation (the different color represent different wavelengths of infrared). Much infrared light is emitted in the plane of the disk by star formation activity in the gases and dust clouds that are concentrated there. The "cloud" of brightness that appears to be above the center of the plane is an area of the constellation Ophiuchus in which stars are being born. The area of Orion (below the plane of the disk, far right), a grouping of stars in the visible-light photograph, emits enormous quantities of infrared light. Orion is seething with star formation activity, with warm, excited clouds of gases and interstellar dust grains throughout. The very active Large and Small Magellanic Clouds, the Milky Way's close companion galaxies, below the plane to the left of Orion, are also strong emitters of infrared.

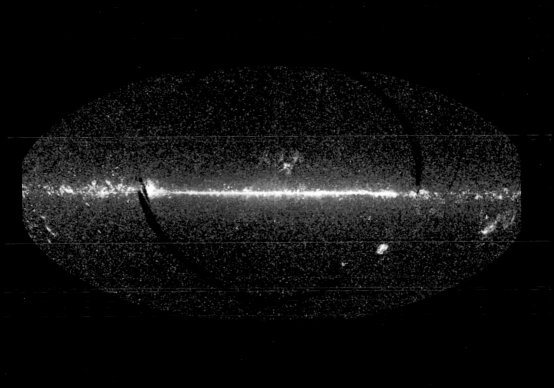

Milky Way Galaxy
View toward the galactic center, in Sagittarius

This visible-light photo looks in the direction of the center of the Milky Way, toward the constellation Sagittarius, the centaur-archer. The Milky Way's center is too far away (28,000 ly), and buried behind too much dust, for us to see it at visible wavelengths. (Most of the brightest stars of Sagittarius are "in front of" the center, just 100 to 200 ly away.) Astronomers have been able to "see" into the center in infrared and radio wavelengths and study the innermost central regions of our galaxy's nucleus. The galactic center is perpetually lighted by the energy from millions of stars. Astronomers are researching the possibility that a supermassive black hole of one to several million solar masses lurks inside a compact radio source known as Sagittarius A, controlling the motions of the stars and gases in this area. Away from the nucleus the density of stars lessens as the central bulge gives way to the disk and spiral arms. Here are found open star clusters, nebulae, and individual stars. Farther out still is the galaxy's halo, inhabited by old, dense globular clusters.

Milky Way Galaxy
View toward Cygnus

This view along the northern Milky Way, taken without telescopic aid, shows the area of the bright Summer Triangle, a trio of stars that marks the northern sky in the warm months. The three stars are bright Vega, in the constellation Lyra, at center-top; Altair, in Aquila, along the bottom of the image to the right of center, with the star beta (β) Aquilae above it; and Deneb, in the constellation Cygnus, the bright star lying left of the bright part of the Milky Way, to the right of the pink hydrogen-emission glow of the North America Nebula, on the far left. This region of the sky is particularly rich in star clouds, bright areas of many distant and unresolved stars, and in dark, light-absorbing clouds of gas and dust. Astronomers think the Cygnus region belongs to the spiral arm in which our solar system resides. Its relative nearness makes it so prominent in our sky. The Sun's galactic orbital motion takes our solar system in the direction of Cygnus, at a rate of 250 km per second (558,000 miles per hour). It takes us about 230 million years to make a complete orbit of our galaxy.

Milky Way Galaxy
View toward Monoceros

This view shows a fainter area of the Milky Way that appears in the southeastern sky during early winter evenings in northern latitudes. The Milky Way runs right through Monoceros, a constellation immediately to the east (left) of Orion, whose belt and sword are visible in this photograph in the upper right. The red supergiant Betelgeuse is at center-top. The noticeable patch of red in the star clouds of the Milky Way is the Rosette Nebula, an important star-forming region. The bright star in the lower left is Procyon, the brightest star in the constellation Canis Minor. There are an estimated 200 billion stars in our galaxy. Under the best of conditions when we look up into the night sky without optical aid we are able to see perhaps a few thousand of the nearest stars, a handful of star clusters and nebulae, and three galaxies beyond the Milky Way. The stars we see lie in a relatively small volume of space in the Milky Way's disk known as the Solar Neighborhood and are typically separated from one another by several light-years.

Horsehead Nebula
Dark nebula in Orion

With its distinctive shape and its location in a rich and beautiful area of Orion, perhaps the most interesting of the constellations, the Horsehead Nebula is one of the most famous nebulae in the sky. It sits just south of Alnitak (zeta [ζ] Orionis), the bright, easternmost star of Orion's belt, shown left of center, but is very difficult to see in amateur telescopes. This image shows the three types of diffuse nebulae. The area of glowing red hydrogen gas, caused by high-energy light from nearby sigma (σ) Orionis, the second brightest star pictured, is an emission nebula. In the lower left is another emission nebula, NGC 2024, lit up by a star hidden in the dark dust lane that partially veils it. The bright blue area below the Horsehead is a reflection nebula, dust illuminated by the light of an embedded star. Finally, the Horsehead Nebula is an extrusion of a dark, or absorption, nebula from a region of thicker nebulosity seen in the bottom right of the photograph. Dark nebulae absorb the light of stars within or near them. Only the stars in front of such nebulae are visible in photographs.

1,500 ly

108

m 4.0

Orion: Sword and Belt Region
Stars and nebulosity in Orion

This visible-light image reveals some of the gases and dust
that contribute to making Orion one of the most active
regions in our sky. The two brightest stars on the left
(north) side are the middle and eastern stars of the belt.
The small, dark intrusion in the glowing pinkish nebulosity
in the lower left is the Horsehead Nebula. The sword, with
the famous Orion Nebula in the center (right side), is lit up
by four, close, bright stars, which excite a large region of
surrounding nebulosity. Nebulae are hallmarks of spiral
galaxies such as the Milky Way. Interstellar material is
gently collected by the galaxy's large-scale gravitational
and magnetic fields into nebulae, clouds of gases and dust.
The intense star-forming activity in Orion is the result of
gravitational density waves formed by the rotation of the
galaxy's disk (these waves also form the spiral patterns).
Material in the disk orbiting the galactic center enters the
spiral arms, where it is slowed and concentrated by the
excess gravity of the concentration of matter already there.
If the material contracts enough to reach a stage called
critical density, it will collapse and star formation will result.

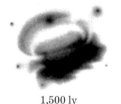

1,500 ly

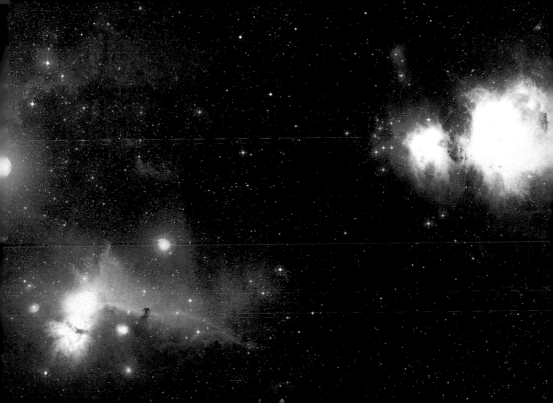

m 4.0, 9.0

M42 and **M43** (NGC 1976 and NGC 1982) Orion Nebula
Emission nebula in Orion

The Great Nebula in Orion, below the belt in the area of
the sword, is the brightest emission nebula in our sky.
Visible even to the naked eye on a clear, dark night as a
fuzzy starlike area, the nebula is quite clear in binoculars
or a small telescope. M42 is the main body of the bright
nebula, lit up by four bright stars; M43 is the smaller
bright splotch to the northeast (upper left). The brightest
stars in the Orion Nebula are the four stars of M42, called
the Trapezium, which are less than a million years old,
quite young in stellar terms. Infrared studies indicate that
star formation is ongoing within the nebula, and in as little
as 1 million years the Trapezium will be joined by more
stars, which will make this region visibly brighter. The
Orion Nebula is the premier location for the study of star
formation. Its distance of approximately 1,500 ly isn't
exactly next door, but considering the scale of activity
going on, it is the nearest rich site of star formation. The
entire constellation is active, in fact, and filled with the
gases and dust (most of it undetectable in visible light)
necessary for star formation to begin.

1,500 ly

M42 (NGC 1976) Orion Nebula
Emission nebula in Orion

This close-up zeroes in on the light source of the Orion Nebula: four stars forming an irregular quadrangle called the Trapezium. These large, hot (hence blue), bright O- and B-type stars formed within the last million years. The most significant of them, with a temperature of 30,000°K, has about 40 times the Sun's mass, is some 15 times its size, and puts out 300,000 times more energy. Even at a distance of 1,500 ly the combined light of these stars is visible to the naked eye. The four stars can be resolved individually with 11-power binoculars. As is typical in beginning clusters, these massive stars will beget new stars by reacting with the gas and dust clouds around them. The stars' radiation and stellar winds push on the surrounding nebulosity, ramming the illuminated areas into a dark nebula behind them, leading to compression and eventually to continuing star formation. The dark areas are not an absence of material but rather a region so dense with dust that light is absorbed; even the light pressure and stellar winds from the powerful Trapezium stars cannot blow the dust away.

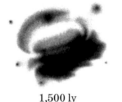

1,500 ly

114

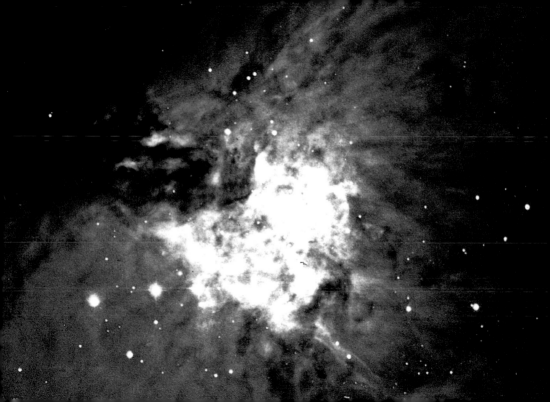

NGC 1973/75/77
Reflection nebula and star cluster in Orion

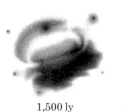

1,500 ly

These very young stars immersed in a series of reflection nebulae are visible as the northernmost point of light marking Orion's sword. More subtle than the turbulent M42 and M43 below it, these objects have a gentle beauty that has generally been overlooked. Most of the light seen in this nebula is of a different origin than that in the Orion Nebula, which is an emission nebula lit by glowing hydrogen gas. The blue in these nebulae comes from the light scattered or reflected by the dust surrounding this small cluster of stars, the brightest of which are blue-hot. The scattering effect makes the nebulosity even bluer than the stars are themselves. Interstellar dust plays a key role in star formation, inhibiting escaping radiation and affecting the rate of formation. You can detect a subtle background of red hydrogen emission in this image as well, as the high-energy light coming from these stars excites the hydrogen atoms, causing them to glow red.

M42 (NGC 1976) Orion Nebula
Protoplanetary disks in emission nebula

1,500 ly

Observation of star formation is made difficult by the great distances of star-forming regions and by the concentrations of dust that obscure certain critical stages. The gravitational collapse of small regions within enormous cool, dark molecular clouds of gases and dust leads to the formation of stars with attendant coplanar planets. When the region collapses to a certain density, gravity begins to round out the dust-thick cloud, and its rotational momentum flattens it to an accretion disk in which planets begin to form around the central "protostar." This unprecedented photograph (obtained by the Hubble Space Telescope after its repair), illuminated by the nebula's ambient light, shows the protoplanetary disks that have formed around several protostars. The protostars in this area, only several hundred thousand to several million years old, shine not by nuclear fusion (they are not full-fledged stars yet) but by the heat generated by their gravitational compression. Radiation and stellar winds from the hot, young star on the left have distorted some of the disks.

118

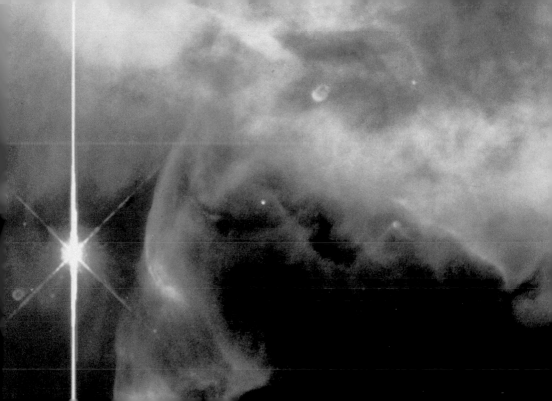

IC 2944
Emission nebula with Bok globules, in Centaurus

370? ly

IC 2944 is an unusually featureless, well spread-out emission nebula that displays the characteristic red glow of excited hydrogen atoms. The dark features silhouetted on the nebulosity are Bok globules, named for Dutch-American astronomer Bart Bok, who recognized these concentrations of gas and dust as the specific sites of star formation. Stars (and their attendant planets) form from large, tenuous interstellar clouds of gas and dust. Over tens of millions of years, the random, swirling motions within the cloud can bring two or more nebular regions into collision, increasing the density to the point at which the resulting gravity collapses the nebula at the area of collision. Once this occurs, the cloud gets smaller and denser. The light-absorbing dust thickens within the shrinking blob, now a Bok globule. Somewhere near the center of the globule, hidden from view, one or more stars are forming. Bok globules are often photographed against a background of glowing gases illuminated by the energy radiated by other recently formed stars.

Beta [β] Pictoris
Type-A star with possible planet formation in Pictor

Beta Pictoris appears to be a routine type-A, 4th-magnitude
star, no more than several hundred million years old, in
the nondescript southern constellation Pictor. But in 1983,
the Infrared Astronomical Satellite (IRAS) detected dust
surrounding and warmed by the star. Follow-up ground-
based observations, with beta Pictoris occulted by a disk
within the telescope, made the first images of a dust disk—
fairly depleted and quite devoid of gases—surrounding a
normal star. The dusty disk, seen almost edge-on, extends
out to about ten times the distance of our outermost planets
from the Sun. The image shown here is a false-color
infrared image that records the heat emitted by the star-
warmed dust. The inner region, the area of which is about
the same as the area of the planets around our Sun, appears
to be particularly devoid of dust. Astronomers speculate
that this dust depletion is due to the presence of planets,
which would have swept up the material in the swift process
of formation, probably within the first 10 million years or
so of the star's life. Discoveries such as this suggest the
possibility that all single stars have attendant planets.

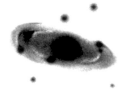

50 ly

122

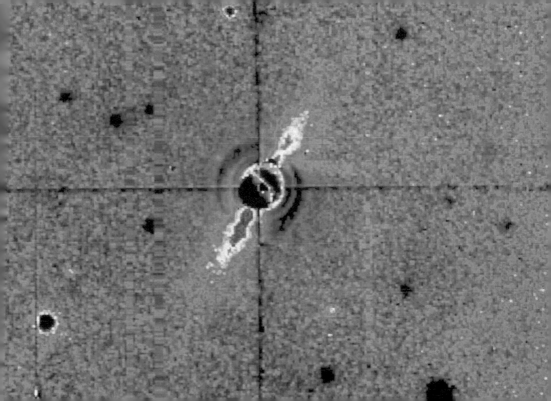

m 6.5

M16 (NGC 3372), Eagle Nebula, with **NGC 6611**
Emission nebula and open cluster in Serpens Cauda

Like the Orion Nebula, the Eagle Nebula is a "stellar
nursery," with a very young star cluster containing several
hot, bright stars and surrounding red nebulosity. The
name comes from the image of the head and wings of a bird
of prey looming across the top. In the center of this bright
nebula is a dark, dense, wing-shaped Bok globule, thick
with light-absorbing dust, which is being distended by
pressure from prevailing winds in the nebula. High-energy
light from nearby stars causes the surface of the tip of
the globule to glow. The predominant red color comes from
hydrogen, the most common element in the universe. When
a hydrogen atom is struck with high-energy light of the
sort (ultraviolet) that comes from hot stars, the electrons
become so energized they jump free of the atomic nucleus.
When the free electrons and nuclei recombine, the
electrons emit light only at the wavelengths (energies)
needed to drop to some particular lower-energy orbit
around the nucleus. At visible wavelengths, the most
common drop results in the emission of light energy at a
wavelength that we perceive as the color red.

6,000 ly

NGC 2264
Emission nebula and open cluster in Monoceros

2,700 ly

The relatively nondescript constellation Monoceros ("one-horned"), which is supposed to represent a unicorn, becomes very interesting and noteworthy through the lens of a telescope. It lies just east of Orion, right along the Milky Way, and contains several beautiful, well-studied star-forming regions. The brightest star in this image is S Monocerotis, a variable star (a star whose light output varies, often periodically) with an average magnitude of 4.7. The region surrounding the star is dominated by a hefty and complex emission nebula, with an associated star cluster, NGC 2264, which is 5 million years old. Dust near some of the hot stars causes some contrasting blue reflection nebulae as well. The dark Cone Nebula (see the following plate) is part of this nebulosity. This great region of active star formation, lying about 2,700 ly from our solar system, is a fine example of how the formation of stars in a cool, dark molecular cloud perpetuates the formation of yet more stars. In such situations an entire cluster of hundreds of stars can form over a period of a few million years.

Cone Nebula
Emission nebula in Monoceros

2,700 ly

This photograph of the Cone Nebula focuses on the turbulent region around the cone for which it is named. This distinctive shape is the result of the surrounding star-formation activities. After stars begin to form from a large, dark, cool molecular cloud, their energy output of both light and material (gas and dust blown out during the final stages of formation) begins to heat and expand into the surrounding cooler and denser regions of the larger nebula. Some of the cooler, denser parts may be Bok globules in the process of collapse, precursors to star formation. These darker, denser regions may retain their integrity, but will yield, like a wind sock, to the hotter stellar winds from the already formed new stars. That is what is going on here in the cone. The lighted end of the cone is not a star itself, but is the excited gases surrounding a newly forming star within. Such clouds of material ejected from new stars, glowing from the intense high-energy radiation from nearby hot stars, are often found in stellar nurseries. They are called Herbig-Haro objects.

128

m 5.2

NGC 2237–39, Rosette Nebula, and NGC 2244
Emission nebula and open cluster in Monoceros

One of the preeminent nebulae in our night sky, extending some 50 to 60 ly across, the Rosette Nebula has long been a favorite object of astrophotographers. Named for its resemblance to a full-blown rose, this nebula is the nursery to the stars in the central cluster, NGC 2244. This is a "brand-new" cluster, with only type-O and type-B stars up to 1 million years old. These are the hot, massive, bright stars that form, live, and die most quickly. Their massive combined outputs of radiation and stellar winds are blowing out the central regions of gas and dust, effectively diminishing star formation there. But as these winds ram into the outer regions of the nebula the resulting increased densities there continue the star formation process. Most of the dark regions in the nebula are probably Bok globules, a preliminary phase in star birth. In a few million years there should be several hundred more stars here, of all representative masses. Despite its large size, the Rosette Nebula has low surface brightness and is difficult to see. NGC 2244, with an overall magnitude of 5.2, is easily visible in binoculars or a small telescope.

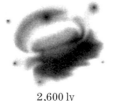

2,600 ly

130

NGC 2237–39 Rosette Nebula
Close-up of central region

The Rosette Nebula lies in a region of the plane of the
Milky Way where the stars appear thickest in the sky, and
most of the stars visible in this view lie between us and
the nebula. The brighter, bluer stars discernable here are
members of the central star cluster, NGC 2244. In this
close-up view of the nebula, more Bok globules are visible.
Over the next 10 million years or so, star-forming activity
will draw to a close as the nebular material either gets used
up in star formation or is blown into interstellar space.
Once the nebula is gone the cluster will remain as an open,
or galactic, cluster. Open clusters contain hundreds to a
thousand or so stars that have formed relatively recently,
during the last few billion years, in the disk of the Milky
Way. Their self-gravity is not strong enough to round out
their shapes, as with the more massive globular clusters.
Further, their identities as clusters are limited by outward
strains induced by the galaxy's general, changing gravity
fields and the shearing effect produced as they orbit the
center of the Milky Way. Eventually their self-gravity
will dissipate, and they will break up.

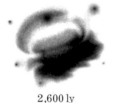

2,600 ly

NGC 2261 Hubble's Variable Nebula
Herbig-Haro object in Monoceros

Hubble's Variable Nebula is a Herbig-Haro object, a thin, glowing globe of dust surrounding an embryonic star. Much of the star formation process takes place behind a thick curtain of dust around the compressing center that eventually grows to become a star and is able to fuse hydrogen into helium to generate its supportive energy. As the process nears completion, and the protostar's attendant planets engage in their coincident formation, the dust clouds surrounding the evolving star begin to thin. They often clear enough that filtered light from the interior shines through, and we can see the unstable protostar—called a T-Tauri star at this brief stage— directly in visible light. Hubble's Variable Nebula has changed in appearance and brightness throughout the decades since its variability was discovered by Edwin Hubble in 1916. While it is possible that the internal protostar may be varying in energy output, it is more likely the dusty nebulosity is variable in thickness, causing the light to dim and brighten as it flows around the protostar, which resides in the bulge to the right.

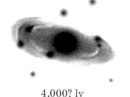

4,000? ly

134

m 5.8

M8 (NGC 6523), Lagoon Nebula, and **NGC 6530**
Emission nebula and open cluster in Sagittarius

The Lagoon Nebula lies in one of the richest regions of the Milky Way, toward the galactic center, not far from the Trifid Nebula, with which it is sometimes photographed. The Lagoon Nebula is larger than either the Great Nebula in Orion or the Rosette Nebula. Its outer dimension is approximately 80 ly in extent, but because of its great distance—3,900 ly—it is not as prominent as its closer peers. (The apparent size of an object depends directly on actual size and inversely on distance.) The Lagoon Nebula displays glowing red hydrogen emission clouds, as well as dark nebulae, where the dust is so thick the light is absorbed. The young star cluster NGC 6530 is apparent on the east (left) side of the nebula. The currently active region of star formation is now to the west, centering on the bright "hourglass" nebula, where the stars are estimated to be only 10,000 years old or less. The region surrounding the Lagoon Nebula appears to be lacking in stars here and there, but this lack is only apparent, caused by widespread absorption nebulae obscuring the stars behind them.

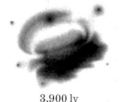

3,900 ly

m 4.5

NGC 2070, Tarantula Nebula, with **30 Doradus**
Emission nebula with open cluster in Dorado

The spider-like Tarantula Nebula is among the largest (800 ly), most massive (5 million solar masses), and brightest emission nebula observed in any galaxy, including our own. It is barely apparent to the naked eye (appearing as a fuzzy star) because of its great distance, 160,000 ly from Earth in our active companion galaxy, the Large Magellanic Cloud. Were the Tarantula Nebula where the Orion Nebula is, at the distance of 1,500 ly, it would dominate the night sky, cover much of the constellation of Orion (an area 60 times the apparent diameter of the Sun), and be bright enough to cast shadows. The presence of such an enormous star-forming nebula in such a small galaxy supports theories that the Large Magellanic Cloud, a generally active galaxy, suffered a recent disturbance, perhaps a pass through or very near our Milky Way. The Tarantula contains many young stars and clusters of stars. It was long assumed that the bright knot at the "body" of the spider was a single, very bright star—recent evidence has proved otherwise (see the next plate, 30 Doradus).

160,000 ly

138

30 Doradus (star cluster R136)
Star cluster in Tarantula Nebula, in Dorado

This phenomenal object has long appeared through earth-bound telescopes as a single, very massive and bright star wrapped in the glow of the Tarantula Nebula. Astronomers studying 30 Doradus in the 1980s concluded that its mass was upward of 2,000 times the mass of our Sun. Established theories on the energy generation of stars indicated an upper limit of no more than 100 solar masses, more likely about 60. Until the recent studies on 30 Doradus, no observationally based star masses contradicted the theoretical work. This incredible mass offered a serious challenge to stellar evolutionary theory, sending scientists scrambling to reconcile the theory with the evidence. The theory won out when evidence, especially from the Hubble Space Telescope, demonstrated that rather than one supermassive star, 30 Doradus was a compact cluster (renamed R136) of at least 3,000 stars, many of which are blue-hot stars. The cluster is remarkably compact, with a huge number of hot, massive stars, and evidently formed with incredible rapidity.

160,000 ly

140

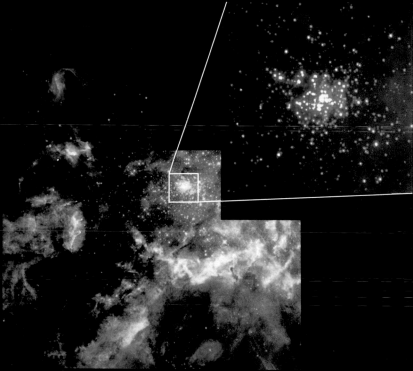

m 8.5

M20 (NGC 6514) Trifid Nebula
Emission nebula in Sagittarius

3,900 ly

Located in the rich part of the Milky Way in the Sagittarius constellation, not far from the Lagoon Nebula, the Trifid Nebula is another favorite subject of astrophotographers. Its aesthetic appeal comes from its manifestation as both a blue reflection nebula, caused by light-reflecting dust around one or more hot stars, and as an emission nebula, whose light comes from excited atoms of various elements, most notably hydrogen, which glows red. The dark dust lanes in the center seem to divide the larger emission nebula in three parts, hence the name Trifid, which means "split in three." This is one of the smaller of our neighboring star-forming areas. Its distance is about 3,900 ly and its extent is about 20 ly. The Lagoon Nebula lies just to the south but is not directly connected to the Trifid, being 900 ly farther from Earth. Although only six stars have been counted in the bright nebulosity of the Trifid, the cloud density is actually typical of stellar nurseries (around 100 atoms per cubic centimeter), and more stars may be hidden or forming in the obscuring dust.

142

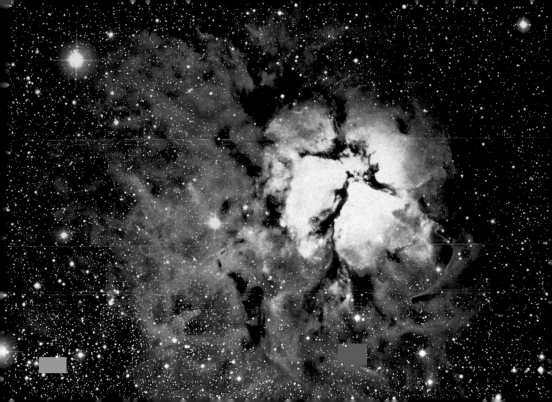

Barnard 86 and NGC 6520
Dark nebula and open cluster in Sagittarius

2,000 ly

This image, taken in the direction of Sagittarius, is filled with foreground stars from the disk of the Milky Way (the small yellow dots). The size of each star reveals the star's apparent brightness, not its size; a brighter star looks bigger in a photograph. Because blue stars don't last long, the typical star is yellow, orange, or orange-red, as in this mass of foreground stars. The larger blue and orange stars here, as well as some of the undistinguished smaller, yellow stars, are members of star cluster NGC 6520. The absence of stars on the right side is the dark nebula Barnard 86. This apparent absence does not indicate a real lack of stars, but rather the presence of a dark nebula, a cloud of gas with enough dust to dim (and redden), sometimes significantly, the light from stars behind it. Although the nebula's dust absorbs the light, only about one particle in a trillion is a dust grain (carbon, silicon, or iron). Virtually all the rest is atoms and molecules of hydrogen and helium. Barnard 86 is most likely the remnant of a larger nebula from which the star cluster NGC 6520 recently formed.

m 9.0

M57 (NGC 6720) Ring Nebula
Planetary nebula in Lyra

The Ring Nebula, a fine example of a class of objects known as planetary nebulae, was the first such object discovered. These objects have nothing to do with planets but were so named because in telescopes they appear as disks, as planets do. Planetary nebulae represent a critical time in a star's life, when it has reached the end of its red giant stage and ejected its outer atmosphere. This material forms a spherical shell that glows when energized by the remaining central star, which is clearly apparent here in the center of the ring. The vast majority of stars will reach the planetary nebulae stage, including billions of years from now, the Sun. (Some very massive stars blow up as supernovas instead, but these are rare.) This stage lasts up to 50,000 years, by which time the ring material has mixed with the interstellar medium. Planetary nebulae are usually too faint for small telescopes, but this one can be discerned as a small gray ball in a 3-inch telescope. As with all celestial objects with low surface brightness, this interesting object is better appreciated with larger-aperture telescopes.

4,000? ly

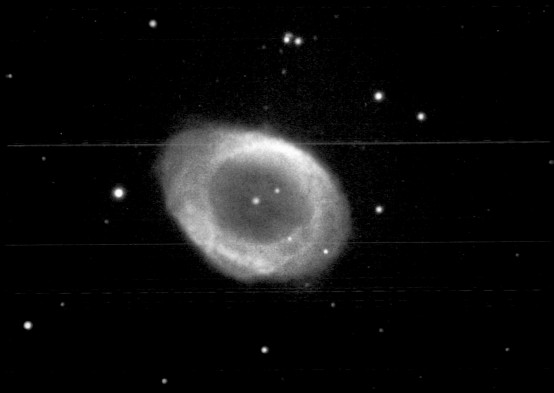

m 6.5

NGC 7293 Helix Nebula
Planetary nebula in Aquarius

Throughout a star's lifetime stellar winds, composed of
charged ions and electrons, stream steadily out into space.
These winds increase as a star evolves. When a star
reaches the red giant stage its larger size lowers the
surface gravity sufficiently to allow the outermost
atmosphere to more easily stream out to space, and the
rate of loss increases. The shell of a planetary nebula is
built up at the end of the red giant stage when stellar
winds increase to their highest rate. This last blast of
atoms, ions, and electrons eventually catches up to the
slower-moving material farther out, at distances typically
ranging from 1,000 to 100,000 times the Earth's distance
from the Sun, and compresses it. Energy released from
this compression and from high-energy, short-wave light
(ultraviolet light and X rays) emanating from the hot
central star, which has been compressing and heating up,
energizes the resulting shell, causing it to fluoresce in the
colors characteristic of the elements present—in the
Helix Nebula, blue-green from oxygen and red and pink
from hydrogen and nitrogen.

450 ly

148

m 8.1

M27 (NGC 6853) Dumbbell Nebula
Planetary nebula in Vulpecula

Planetary nebulae do not always present themselves
as a neat, glowing, expanding shell, as seen in the Ring
Nebula. The variety of shapes, some quite complex, in
planetary nebulae have led astronomers to look into the
later stages of the stellar winds that red giants exude.
Slower stellar winds blow prior to the rapid blast at the
end; the earlier winds stream into space more easily near
the equatorial plane of the star. The buildup of ions and
electrons in the equatorial plane restricts the later flow
of wind in that plane. When the rapid ejection occurs
at the end, the winds meet less resistance toward the
rotational poles of the star, resulting in the formation
of two large lobes at the pole ends, giving the nebula
an hourglass shape. The shapes we see are primarily a
function of perspective. If seen from the pole end, this
hourglass-shaped structure will look like the Ring or the
Helix nebulae. Those seen at other angles will display
more complexity. Evidently we view the Dumbbell
Nebula on or near the plane of its equator, so the two-
lobed shape is apparent.

3,500 ly

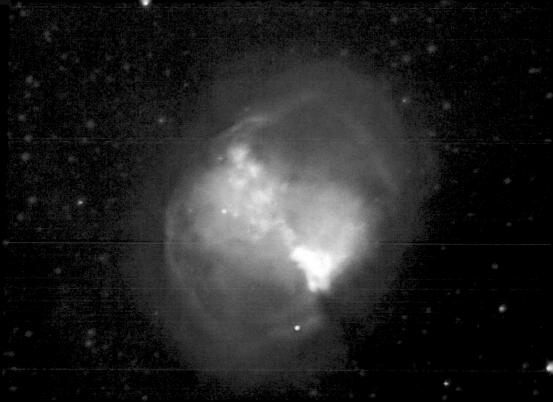

NGC 2440
Planetary nebula in Puppis

6,500 ly

The mass ejection of a red giant's outer material occurs when the nuclear fuel, usually helium or carbon, is used up, shutting down the nuclear furnace within the core. A star can eject from 10 to 60 percent of its mass—and significantly more of its volume, because it sheds its least compressed material. This loss allows gravity to compress and heat the star to extremes, driving off the atmosphere at a furious clip until a level is reached at which the surface gravity becomes strong enough to restrain further outgassing. The surface of the central star, newly exposed "skin" that had been buried deep within the star, is extremely hot. NGC 2440's central star is the hottest star known, with a surface temperature of at least 200,000°K (360,000°F). The nebula around it has no clear symmetry, perhaps due to a violent expulsion or distortion by interstellar material. Eventually, when the central star of a planetary nebula is about the size of Earth, gravity can compress it no further; at this stage it is a white dwarf. It then cools over a span of billions of years until it is as cold and black as space itself.

152

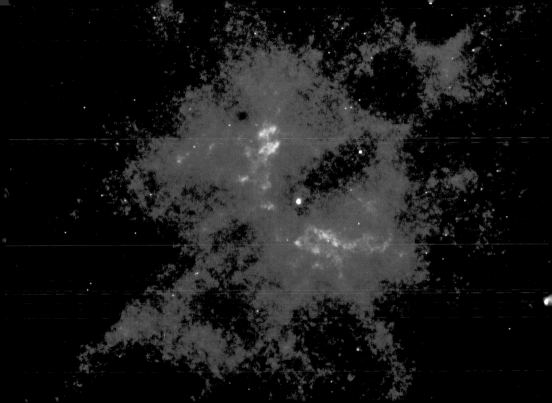

m 8.4

M1 (NGC 1952) Crab Nebula
Supernova remnant in Taurus

In 1054 there appeared in the sky a star so bright it was
visible in daylight for more than a month. Over a two-year
period it faded slowly to invisibility. Such stars, found
occasionally in historical records, became known as *novas*,
Latin for "new." Actually old stars, some of them nearing
the end of their lives, they are now classified as *novas*,
white dwarfs that experience surface explosions caused by
hydrogen blowing off a close giant star; and *supernovas*,
categorized in two groups. Type I supernovas occur when
hydrogen from an expanding giant star blasts a white
dwarf companion, causing it to explode entirely. Type II
are very massive, unstable stars that explode at the end
of the nuclear fusion stage of their lives; afterward a
huge, irregular shock wave composed of all the known
elements—called a supernova remnant (SNR)—is blasted
into space. The Crab Nebula is an SNR from a Type II
supernova. Measurements of its rate of expansion
suggest that this SNR, now about 4.4 ly across, started
its expansion about 950 years ago, and that it was the
nova recorded in 1054.

6,500 ly

154

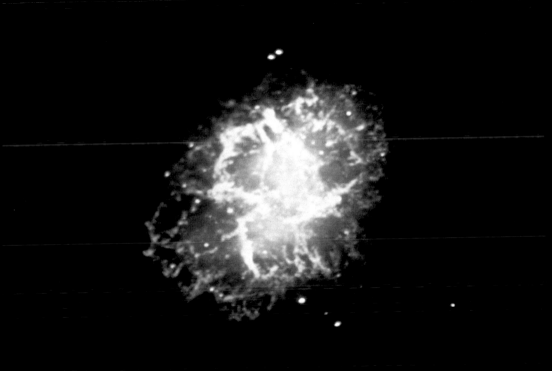

Gum Nebula
Supernova remnant in southern skies

When a massive star reaches its final "energy crisis," it immediately loses a large amount of energy in its core, which implodes by rapid gravitational collapse. This rapid squeezing leads to the generation of incredible quantities of neutrinos (uncharged elementary particles believed to have no mass) and a violent explosion (such stars may be 20 times more massive than the Sun). A supernova puts out more light than all the stars within its galaxy (hundreds of billions of stars) combined. The Gum Nebula the nearest supernova remnant (SNR) to Earth, is so large (extending 2,300 ly) and close (the nearest portions are about 300 ly away) that it spreads out over 60° of the southern skies. Until the 1950s, when an astronomer named Colin Gum studied it, it was thought to be many isolated nebulae. The supernova from which it came exploded about 11,000 years ago and would have rivaled the Moon in brightness. The star is now a pulsar in the constellation Vela, about 1,300 ly from Earth. The SNR's shock wave, composed of high-speed ions and electrons from the blast, is still expanding in our direction.

300 ly

Supernova 1987A
Supernova in Dorado

Astronomers see supernovas every year, usually in distant galaxies, after the explosion's peak. When SN 1987A, the first supernova visible to the naked eye since 1604, blew up in our companion galaxy the Large Magellanic Cloud in 1987, astronomers caught it in the act. Such explosions occur when the core of a massive star that can no longer produce energy by nuclear fusion, crushed inward by gravity, implodes. In an instant the Earth-sized core collapses to a neutron star of unimaginable density, only 25 km (15 miles) in diameter. Most of the energy of the implosion escapes the star as neutrinos, massless, chargeless particles. The neutrinos from SN 1987A took 169,000 years to reach Earth, probably a total of 100 billion per square inch; only 19 were detected by Earth-based sensors. Among the elements supernovas can generate are lead, gold, and uranium. This image shows SN 1987A before, as a blue supergiant (type B3), and after the blowup. Initially expanding at the phenomenal rate of one-tenth the speed of light, SN 1987A faded rapidly—but its legacy will keep astronomers occupied for years.

169,000 ly

Supernova 1987A: Aftermath
Supernova remnant in Dorado

169,000 ly

Shortly after the explosion in 1987, astronomers detected a bright, small ring of material centered on SN 1987A's original star composed of elements—carbon, nitrogen, and oxygen—expelled by the star over the last 400,000 years of its life. Seven years later the Hubble Space Telescope confirmed the existence of two mysterious, very large, symmetrical rings extending several light-years into space. They are not centered on the supernova but may lie along the polar axis of the original star. Astronomers speculate that the radiation that makes the rings glow is in the form of two oppositely directed beams coming from the location of the explosion, swinging around and illuminating the material that was blown out, in the standard hourglass shape, by the stellar winds when the star was in the red giant stage. This would provide evidence that a pulsar, a rapidly rotating star that sends out beams of energy, remains from the star that exploded. The rings are actually circular, though they look elliptical due to perspective. The two bright objects in the upper left and lower right are other stars.

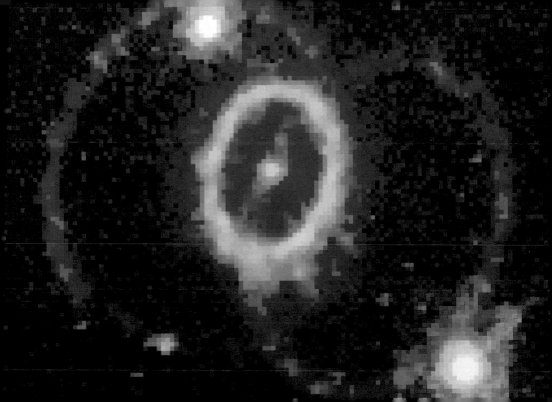

m 1.2

M45 Pleiades
Open cluster in Taurus

One of the most famous objects in the sky, the Pleiades h
been attracting the attention of humans since prehistoric
times. It lies on the back of Taurus, the Bull, in the winte
evening skies. It is also known as the Seven Sisters, becau
six or seven stars are visible to the naked eye (some myth
account for the missing seventh sister). Binoculars revea
hundreds of fainter stars and thereby the true nature of
this object—a young, compact star cluster (only about
100 million years old), 400 ly away, composed of several
hundred stars in a region about a dozen light-years acros
Dust particles reflecting light from the stars give the
cluster its bluish glow in photographs. As one of the
brightest open clusters in the sky, the Pleiades has
contributed much to the understanding of star clusters.
Open clusters are born in the planes of galaxies, where
they are subject to gravitational interference from outsic
forces. With their sparse populations, they do not have
the self-gravity to hold together indefinitely against this
onslaught, and most eventually (within 1 or 2 billion year
dissipate into the surrounding galaxy.

400 ly

162

m 4.4

NGC 869 and NGC 884 h and χ Persei
Open clusters in Perseus

The stars of this attractive pair of open clusters, also known as the Perseus Double Cluster, can be resolved in binoculars or a small telescope. Most of the brightest stars are blue, and the other bright ones are red giants, a signature of young clusters less than 500 million years old. The stars of these relatively shapeless clusters tend to be richer in heavy elements (though they are mainly hydrogen and helium) than those of older globular clusters because they formed from the debris left by the previous generation of stars. These earlier stars, called first generation, are pure hydrogen and helium, virtually the only elements that existed after the Big Bang, but late in life they convert helium to heavier elements (such as oxygen, carbon, and even iron and uranium), which are expelled into space when the stars die. Most open clusters are found within the planes of spiral galaxies (this is why they are also called galactic clusters). In the Milky Way they are found in the cloud-laden spiral arms, where most star formation takes place.

7,000 ly

164

m 5.0

NGC 4755 Jewel Box
Open cluster in Crux

Lying along the Milky Way, the constellation Crux, the Southern Cross, contains in its small area at least ten open clusters visible with small telescopes. NGC 4755 is by far the brightest of these and one of the prettiest clusters in the sky. It is sometimes called the Kappa (κ) Crucis Cluster as it is superimposed on that star. Located just southeast of beta (β) Crucis, at a distance of some 7,800 ly from Earth, it consists of about 50 stars ranging in magnitude from 6.0 to 12.0 (most are on the fainter side), spread over an area 10′ wide. NGC 4755 is of great astrophysical interest because it is among the brightest and youngest clusters known. Most of the stars are bright blue and yellow-orange supergiants probably not more than 10 million years old. The stars in a cluster form at about the same time from the same interstellar material, and so are about the same age, of similar composition, and about equidistant from Earth. Even though their gravitational forces are weak compared to those of globular clusters, the stars in open clusters are gravitationally bound together and travel through space in the same direction and at about the same speed.

7,800 ly

m 6.9

M67 (NGC 2682)
Open cluster in Cancer

M67 is an open cluster of 500 stars spread out over an area of 12 ly, with an apparent angular size of about 15', half the full moon's diameter, and an overall magnitude of 6.0. The atypical characteristic of this cluster is its age of 5 billion years. Open clusters generally are known for their amorphous shapes, youth, and proximity to the galactic plane. Because they tend to have weak gravitational forces they do not hold a shape the way globular clusters do. Positioned in or near the galactic plane, they experience much tugging and pulling from the large numbers of other stars near them. Their self-gravity is rivaled by these outside forces, and over time these clusters spread out and dissipate into the disk of the Milky Way. Most open clusters are less than 2 billion years old. Our Sun, for example, travels alone in space today, but when it formed 4.5 billion years ago, it did so in the company of hundreds of other stars, long since gone their separate ways. The extreme age of M67 may be explained by its location. At a distance from Earth of 2,600 ly, it lies outside of the Milky Way's disk and thereby experiences less outside interference.

2,600 ly

Hyades
Open cluster in Taurus

Prominent in winter evening skies, this cluster, shaped like the letter V, marks the face of Taurus, the Bull. Although part of the V, the bright red giant Aldebaran (alpha [α] Tauri), is not a member of the cluster; it is about 80 ly closer to Earth. The Hyades cluster contains about 200 half-billion-year-old stars in an area 6° across. The nearest sizable cluster to us (140 ly), it is the most easily studied. It is close enough for its motion to be detectable, allowing astronomers to use a fundamental geometric method (not based on assumptions or the calibration of some other object) to determine its distance, called the moving cluster (or converging point) method. Although the individual stars have their own peculiar motions, they share an overall parallel motion toward a particular point in the sky, but their motions don't appear parallel because they are converging on a distant point. The geometry of their motion vectors enables us to determine their distance accurately. The cluster can then be used to test and calibrate other methods that can be applied to greater distance scales, even intergalactic ones.

140 ly

170

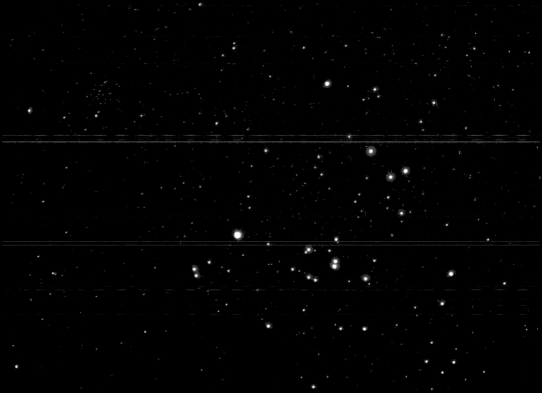

m 3.9

M44 (NGC 2632) Praesepe Cluster
Open cluster in Cancer

In the Solar Neighborhood, the region within several
hundred light-years of our solar system that generally
shares in the Sun's orbital motion around the galaxy's
center, there are several notable open star clusters,
including the Pleiades and the Hyades, both in Taurus,
and the Praesepe, or the Beehive, shown here. The most
distant of the three, at 520 ly, it is the fifth-largest
appearing cluster in our sky, with a diameter of about
80', almost three times that of our Moon. Try the averted
vision technique when you've spotted this rather faint
object: Focus your eyes right next to the cluster to
make this fuzzy patch quite visible. With an age of
approximately 800 million years, the Praesepe Cluster

520 ly

will probably have dissipated into the Milky Way's disk
within the next ten solar orbits of our galaxy (2–2.5 billio
years). Many of its stars will remain but will have been
scattered by the gravitational tugging of the galaxy. This
will also be the fate of our other cluster neighbors, the
Hyades and the Pleiades, although the younger Pleiades
may last a bit longer.

m 3.7

NGC 5139 Omega Centauri
Globular cluster in Centaurus

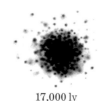

17,000 ly

Globular clusters are far older and more massive than open clusters. Rather than hundreds of stars they contain hundreds of thousands, even millions. With so much more mass, their gravitational forces are sufficient to round them out into a spherical distribution. Most globular clusters were formed early in the evolution of their parent galaxy, and they tend to be found in the halo of the galaxy rather than within its plane. The location, age, and chemical composition of the globular clusters of the Milky Way are of great interest to astronomers, who study their properties to learn about the origin and early aeons of our galaxy, as well as the evolution of stars of the early universe, which generally lack elements higher in the periodic chart than helium. Omega Centauri is visible to the naked eye in the southern skies and is, in fact, the most easily viewed globular cluster. It lies at a distance of 17,000 ly and contains more than 1 million stars within its 320-ly diameter. The stars are so crowded near the center they may be only 0.1 ly from one another (our Sun is 4.3 ly from alpha [α] Centauri, the nearest star).

m 4.0

NGC 104 (47 Tucanae)
Globular cluster in Tucana

Most globular clusters contain only old, first- or second-generation stars that formed from interstellar matter ear[?] in the universe and lack elements heavier than hydrogen and helium. The globular clusters of the Milky Way, most of which reside in the halo, beyond the plane of the disk, contain some of the oldest stars in our galaxy. Globulars usually do not have old stars mixed with more recently formed stars, as is the case with clusters in the Milky Way disk, nor do they have much in the way of interstellar gas and dust, necessary for star formation. It is probably because of their isolation from the relatively busy and crowded disk that the clusters do not have access to star-forming materials. But their location—far removed from interfering gravitational forces that can pull open cluster apart—has enabled them to maintain their spheroidal shapes over the lifetime of our galaxy (14–16 billion years). NGC 104, long considered a single star called 47 Tucanae, is the second largest and brightest globular cluster in the sky, after Omega Centauri. Its angular size 23′ across, is almost as large as that of the full moon.

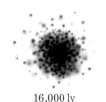

16,000 ly

NGC 104 (47 Tucanae)
Close-up of core region

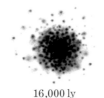

16,000 ly

The Hubble Space Telescope recorded this image in 1994, shortly after it was repaired. It shows pinpoint images of never-before-resolved stars in the direction of the densely populated core of this globular cluster of more than 1 million stars. Such images allow astronomers to make accurate measures of star magnitudes and colors, critical for determining the properties and evolutionary states of such clusters. One surprise yielded by this research was the unprecedented finding of white dwarfs in a globular cluster. This is stunning, considering this cluster's great distance of 16,000 ly and the intrinsic faintness of these hot but tiny stars, typically 5,000 to 10,000 times less luminous than the Sun. Since the colors of white dwarfs are a function of their age, these observations will offer astronomers a new means of determining the age of this cluster, which relates to the age of our entire galaxy. With the greater resolution now available, astronomers can begin to study the interiors of globular clusters, which, although they contain no potential for new star life, may hold keys to the galactic past.

m 6.0

M5 (NGC 5904)
Globular cluster in Serpens

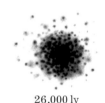

26,000 ly

Some globular clusters are the oldest collections of stars in the universe. Active star-forming regions give birth to stars of blue, white, yellow, orange, and red coloration, but the blue stars, being hot and large, outshine the others. The yellowish overall color of globular clusters results from a lack of young blue stars. These clusters do not have the raw materials—gases and dust—needed for star formation. Studies of their stars' chemical composition indicate that after the globular clusters formed at the beginning of the evolution of our galaxy, star formation within them soon died off. The gases and dust that must have been present were somehow wiped out. Astronomers found the answer in the orbits of these clusters. Their orbits are inclined (tilted) often highly so, to the Milky Way's disk. When their orbital paths took them through the young Milky Way's disk, the material of the disk drew the material out of the globulars, effectively bringing to an early end any star-formation potential. Due to its relatively high brightness and northerly position, M5 is one of the more easily observed globular clusters for North American skywatchers.

Observing Deep-Sky Objects

Although most of the photographs in this book were made with powerful professional equipment, the amateur observer, too, can observe deep-sky objects. You will not be able to achieve the clarity and detail seen in most of these photographs, but you can see many fascinating and wonderful things on a clear, dark night with a pair of binoculars or a small or moderate-size telescope.

Skywatching Conditions

The quality of our view of the sky is greatly affected by the atmosphere. When we gaze into the sky from Earth's surface we must look through miles of turbulent, filtering gases. When observing at lower elevations (near the horizon), we look at a greater slant through the atmosphere, hence through more of it, and the interference is increased. The atmospheric gases have three major effects on what we see. First, they affect *transparency*—the clearness of the sky—making celestial objects appear dimmer than they would in the absence of atmosphere. Second, the atmosphere filters light differently at different wavelengths. Visible light passing through the atmosphere is reduced, through scattering, noticeably more in the blue region of the spectrum than the red. The greater the transparency, the less the reddening effect of the atmosphere. Third, the turbulence of the air blurs what we see because the light beams heading toward us are bent slightly and variably by moving air pockets of differing densities, causing an object in the sky to appear to blur or jump around slightly. Crucial elements for observing deep-sky objects, which tend to have a low surface brightness, are a clear night and a dark location. City lights or a large, bright Moon can cause too much light pollution for this subtle work. Once you are outside, it will take your eyes a few minutes to adapt to the darkness.

Locating Deep-Sky Objects

Certain deep-sky objects are fairly easy to find, such as the Orion Nebula, found below Orion's belt in the area of the sword. We have occasionally given angular measurements between objects to help in locating them. These measurements, explained in "Systems of Measurement," are expressed in degrees (°), minutes of arc (′), and seconds of arc (″). For in-depth viewing, good sky maps or individual constellation charts, such as those in the *ational Audubon Society Pocket uide to Constellations of the orthern Skies* and the *National udubon Society Field Guide to e Night Sky,* are essential for cating galaxies, nebulae, and ar clusters in the sky.

nother thing to keep in mind hen skywatching is that different arts of the sky come into view t different times of year. As arth turns, an observer at any articular location sees different arts of the celestial sphere in his r her line of sight. Every day e rotate 360° to the east; every our, 15°; every 4 minutes, 1°. s seen from Earth, objects in he sky appear to move from east o west at this rate. Every night he stars rise 4 minutes earlier han they did the night before this is because of Earth's movement along its orbit). Over the course of a month,

about 30 days, stars will be rising 2 hours earlier (4 minutes/day × 30 days = 120 minutes).

Observing Instruments

There are many types of telescopes and binoculars on the market that are geared to amateur skywatchers. The advantages of such instruments are that they are able to collect more light than the human eye and they can magnify objects, allowing you to study the surfaces of the Moon and planets and to view many deep-sky objects in detail.

Binoculars

A good pair of binoculars of at least 50 mm aperture can greatly increase your viewing capabilities. Binoculars come with different magnification, or power, capabilities and in different objective lens diameters, or apertures. The greater the

magnification for a given aperture, the narrower the field of vision and the lower the contrast between light and dark. The larger the aperture of the lens the better, because larger apertures allow more light in. The ideal binoculars for the amateur skywatcher are those designated 7×50, with a magnification of 7, meaning objects are magnified seven times, and a 50 mm aperture. Such binoculars will allow you to see features on the Moon, to explore the star clouds of the Milky Way, and, if held steadily, to observe the Galilean satellites of Jupiter. Binoculars that are too powerful (over $8\times$ magnification) are difficult to use unless they are mounted on a tripod or other mount. For long-term viewing a mounting device that allows you to use the binoculars with a camera tripod is recommended. With the binoculars securely mounted, you can see more easily and you won't tire your arms.

Telescopes

If you are certain of your interest in astronomy, you may wish to purchase a telescope. The advantages of telescopes over binoculars are that they gather much more light, have far greater powers of magnification, and come with a mounting system to hold them steady. It is highly recommended that you buy a good telescope, which will cost at least $500, unless you get a good deal on a used one. The relatively inexpensive telescopes sold in department stores are usually of poor quality and will be difficult and discouraging to use. Good-quality scopes are sold at telescope shops, planetariums, and by mail-order firms that advertise in astronomy magazines. You can sometimes purchase a used telescope from a member of a local astronomy club.

There are three basic types of amateur telescopes. *Reflector* telescopes use a system of mirrors to collect and reflect light to the point of focus. *Refractor* scopes use a lens to collect and bend (refract) light to the focal point, as in binoculars. *Catadioptric* systems employ both a lens and a mirror. In all three types an eyepiece, or ocular, at the focal point magnifies the image. Reflector telescopes tend to be less costly than refractors and are generally more compact. For a given aperture refractors are more expensive than reflectors but are sturdier and less likely to be knocked out of alignment. Catadioptric telescopes, such as the Maksutov and Schmidt-Cassegrain systems, are more expensive than either reflectors or refractors but often perform better. They are also usually more compact and therefore easier to transport and use.

The size of a telescope refers to the diameter of the main light-gathering element (a mirror in reflector scopes, a lens in refractors). A telescope whose main light-gathering element (or objective) is less than about 8 inches in diameter is considered a small instrument. Telescopes with objectives of about 8 to 16 inches are referred to as moderate size for amateur use. Anything larger than 16 inches is considered large in amateur terms.

Three variables affect the quality of the image you see in

telescope: the objective's diameter, its focal length (the distance from the primary objective and the focal point); and the ratio of the diameter to the focal length. The larger the diameter of the objective, the greater the light-collecting power and resolution; the longer the focal length, the narrower the field of view and the higher the possible magnification. Telescopes with low focal ratios (the focal length divided by the diameter) have wide fields of view but lower magnification powers. Those with higher focal ratios are longer and more difficult to use but allow for higher magnification. High-power magnification is not always desirable. Higher powers yield a smaller field of view and are harder to aim, focus, and keep steady. Most observing is done at powers of less than 100.

You will want to use lower powers for viewing the Moon and star clusters and for scanning wide fields.

Mounts

There are two basic ways to mount a telescope. The simplest, called an *altazimuth mount*, resembles a camera tripod and allows the telescope to move horizontally (in azimuth) and vertically (in altitude). The problem is that as Earth rotates, celestial objects continually change both azimuth and altitude, which means constant adjustment of both axes of motion in order to track an object. The more expensive *equatorial mount* is oriented to the polar axis, and can be moved north-south and east-west. Once you point the telescope to an object, you can lock the north-south axis, and move the east-west axis to keep the object in view (a motor drive can do this for you).

Systems of Measurement

Many of our modern systems of measurement were derived from methods developed by ancient skywatchers to specify the locations of objects in the sky. Astronomers use systems of measurement to gauge brightnesses, sizes, positions, and distances of celestial objects.

Units of Measure

The basic unit of measurement used for celestial distances is the *light-year* (ly), the distance that light travels in a vacuum in one year. Since light travels at a rate of 299,792.458 km per second (186,282 miles per second), in one year it travels 9.46×10^{12} km (5.88×10^{12} miles). An *astronomical unit* (a.u.) is the measurement of Earth's average distance from the Sun—149,597,870 km (about 92,960,000 miles). One light-year is 63,240 a.u. A *solar mass* refers to the mass of the Sun, which is 332,946 times Earth's mass.

Magnitude

An object's brightness in our sky is called its apparent magnitude. About 2,000 years ago the Greek astronomer Hipparchus designated the brightest stars in the sky "1st magnitude" and those just barely visible to the unaided eye "6th magnitude." This scale was later quantified more precisely so that each step in magnitude equals a factor in brightness of 2.512 times. A star of 1st magnitude is therefore almost exactly 100 times (2.512^5) brighter than a star of 6th magnitude. Some objects are brighter than 1st magnitude; their apparent magnitudes are expressed in negative numbers. The brightest star in the sky, Sirius (α Canis Majoris) has a magnitude of −1.46, and Venus, at its brightest, about −4. The Sun's apparent magnitude is about −27. The apparent magnitude of an object depends on both its intrinsic brightness and its distance from Earth. An intrinsically faint star, if close to Earth, can appear brighter than an intrinsically bright star that is much more distant.

The Celestial Sphere

Astronomers have also developed coordinate systems (similar to terrestrial latitude and longitude systems) for plotting objects in the sky and gauging distances between them. Some coordinate systems are fixed on the *celestial sphere*, the "dome" of the heavens. The celestial sphere is an infinitely large imaginary sphere surrounding Earth onto which are "pasted" all the objects seen in the

The Celestial Sphere and Equatorial Coordinates

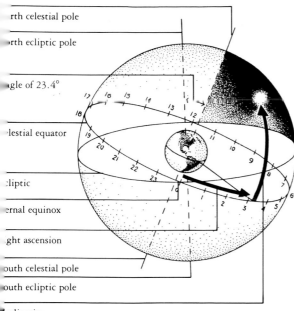

- rth celestial pole
- rth ecliptic pole
- gle of 23.4°
- lestial equator
- :liptic
- ernal equinox
- ght ascension
- outh celestial pole
- outh ecliptic pole
- leclination

sky. The *celestial equator* is a projection of Earth's equator onto the celestial sphere. Similarly, the *celestial poles* are projections of Earth's north and south geographic poles onto the celestial sphere. The *ecliptic* can be seen as the projection of Earth's orbit onto the celestial sphere or, in terms of the Sun, as the path traced by the Sun around the sky (this path is inclined at 23.4° to the celestial equator because of Earth's 23.4° tilt). The *ecliptic poles* are perpendicular to the plane of the ecliptic. The directions "north" and "south" in the celestial sphere mean toward the north and south celestial poles, respectively. "East" always means toward your eastern horizon.

Equatorial Coordinates

One commonly used set of celestial coordinates, called

equatorial coordinates, is based on the celestial equator. The entire celestial sphere is divided into 360°. One degree can be divided into 60 *minutes of arc* (60′), and one minute of arc can be divided into 60 *seconds of arc* (60″). The north-south coordinate, the celestial equivalent of latitude on Earth, is called *declination*. It is measured, like latitude, in degrees, minutes of arc, and seconds of arc, from 0° at the celestial equator to 90° north and 90° south at the celestial poles. The east-west coordinate, the celestial equivalent of longitude on Earth, is called *right ascension;* it is usually measured eastward around the sky in hours (^{h}), minutes (^{m}), and seconds (^{s}) of time but can also be measured in degrees. Since Earth rotates 360° in 24 hours, one hour (1^h) of right ascension, or time, equals 15° of arc; one minute (1^m) of right ascension equals 15 minutes (15′) of arc; and one second (1^s) of right ascension equals 15 seconds (15″) of arc. The "zero point" of right ascension, celestial kin to Earth's Greenwich meridian of longitude, is the point at which the Sun (and thus the ecliptic) crosses the celestial equator on its way north each spring; this point is called the *vernal equinox.*

The equatorial coordinate system is the one used most frequently to specify the positions of fixed celestial objects. Catalogs of stars and of deep-sky objects specify the right ascensions and declinations of these objects at some particular time, or epoch. These coordinates change slowly over time because of a gradual change in the orientation of Earth's axis (precession).

Angular Size

The division of the celestial sphere into 360° not only gives us coordinates with which to locate objects but also provides a unit of measure. You can use your fist to measure the approximate angular sizes of objects on the celestial sphere or the distance between two objects in degrees, minutes of arc, and seconds of arc. Held at arm's length your fist is about 10° across at the knuckles. The tip of your index finger measures about 2°. Viewed from Earth, the Sun and Moon are each ½° across (or 30′, 1,800″, or 2^m). If you read in the text that an object is 4° north of another, you can use your fist to measure the distance and locate the object.

Credits

Photographers
© Anglo-Australian Observatory, photography by David Malin (27, 29, 51, 55, 59, 61, 77, 113, 115, 117, 121, 125, 127, 129, 133, 139, 143, 145, 149, 159, 177, 181))
Dr. Reginald Dufour (53, 65)
Richard Hill (105, 107, 171, 173)
Jet Propulsion Lab (39, 101, 123)
Lund Observatory (99)
Courtesy NASA (97, 153, 161)
National Optical Astronomy Observatories (23, 25, 35, 37, 43, 47, 49, 57, 67, 69, 71, 73, 75, 79, 85, 87, 89, 93, 95, 103, 135, 147, 151, 155, 165, 167, 169, 175)
© Royal Observatory Edinburgh/ Anglo-Australian Observatory, photography by David Malin (63, 91, 109, 111, 131, 137, 157, 163)
Space Telescope Science Institute (31, 33, 41, 45, 81, 83, 119, 141, 179)

Cover Photograph: Spiral galaxy M83, by Dr. Reginald Dufour
Title Page: Horsehead Nebula, © Anglo-Australian Observatory, photography by David Malin
Pages 20–21: Stephan's Quintet galaxy cluster, National Optical Astronomy Observatories

This book was created by Chanticleer Press. All editorial inquiries should be addressed to:
Chanticleer Press
568 Broadway, #1005A
New York, NY 10012
(212) 941-1522

To purchase this book, or other National Audubon Society illustrated nature books, please contact:
Alfred A. Knopf, Inc.
201 East 50th Street
New York, NY 10022
(800) 733-3000

Chanticleer Press Staf

Publisher: Andrew Stewart
Managing Editor: Edie Locke
Art Director: Amanda Wilson
Production Manager: Susan Schoenfeld
Photo Editor: Giema Tsakuginow
Photo Assistant: Consuelo Tiffany Lee
Publishing Assistant: Alicia Mills
Text Editor: Amy K. Hughes
Copyeditor: Patricia Fogarty
Natural Science Consultant: Richard Keen
Picture Editor: Alexandra Truitt
Picture Researcher: Jerry Marshall
Illustrations: Acme Design, Ed Lam (p. 187)
Original series design by Massimo Vignelli

Founding Publisher: Paul Steiner